Math

ADVANTAGE

HARCOURT
BRACE

Orlando • Atlanta • Austin • Boston • San Francisco • Chicago • Dallas • New York • Toronto • London

http://www.hbschool.com

Printed in the United States of America

ISBN 0-15-310691-3

7 8 9 10 030 02 01

Senior Authors

Grace M. Burton
Chair, Department of Curricular Studies
Professor, School of Education
University of North Carolina at Wilmington
Wilmington, North Carolina

Evan M. Maletsky
Professor of Mathematics
Montclair State University
Upper Montclair, New Jersey

Authors

George W. Bright
Professor of Mathematics Education
The University of North Carolina at Greensboro
Greensboro, North Carolina

Sonia M. Helton
Professor of Childhood Education
Coordinator, College of Education
University of South Florida
St. Petersburg, Florida

Loye Y. (Mickey) Hollis
Professor of Mathematics Education
Director of Teacher Education and Under-
  graduate Programs
University of Houston
Houston, Texas

Howard C. Johnson
Dean of the Graduate School
Associate Vice Chancellor for Academic Affairs
Professor, Mathematics and
  Mathematics Education
Syracuse University
Syracuse, New York

Joyce C. McLeod
Visiting Professor
Rollins College
Winter Park, Florida

Evelyn M. Neufeld
Professor, College of Education
San Jose State University
San Jose, California

Vicki Newman
Classroom Teacher
McGaugh Elementary School
Los Alamitos Unified School District
Seal Beach, California

Terence H. Perciante
Professor of Mathematics
Wheaton College
Wheaton, Illinois

Karen A. Schultz
Associate Dean and Director of Graduate Studies
  and Research
Research Professor, Mathematics Education
College of Education
Georgia State University
Atlanta, Georgia

Muriel Burger Thatcher
Independent Mathematics Consultant
Mathematical Encounters
Pine Knoll Shores, North Carolina

Advisors

Anne R. Biggins
Speech-Language Pathologist
Fairfax County Public Schools
Fairfax, Virginia

Carolyn Gambrel
Learning Disabilities Teacher
Fairfax County Public Schools
Fairfax, Virginia

Lois Harrison-Jones
Education Consultant
Dallas, Texas

Asa G. Hilliard, III
Fuller E. Callaway Professor
  of Urban Education
Georgia State University
Atlanta, Georgia

Marsha W. Lilly
Secondary Mathematics
  Coordinator
Alief Independent School District
Alief, Texas

Judith Mayne Wallis
Elementary Language Arts/
  Social Studies/Gifted Coordinator
Alief Independent School District
Houston, Texas

CONTENTS

CHAPTER 1
Sorting and Classifying

Circle the buttons that look like the button in the middle.

Photography Credits:

Harcourt Brace & Company Photographs

Key: (t) top, (b) bottom, (l) left, (r) right, (c) center.

Cover, 2 Weronica Ankarorn; 3 (br) Victoria Bowen; 4, 5, 6, 11,12, 13, 14, 15 (insert) Bartlett Digital Photography; 15 (t), (br), 16 (t), (b) Rich Franco; 16 (insert) Bartlett Digital Photography.

Ilustration Credits:

Pam Perkins: Cover, 2.

Harcourt Brace School Publishers

Dear Family,
Today we started Chapter 1. We will be learning to put together things that are the same in some way. Here and in the Home Notes are some things you can do to help me learn math.

Love,

_____

Talking Math

Use these math words as you talk with your child about his or her work.

alike (the same in some way) These shapes are alike because both are red.

not alike (different in some way) These shapes are not alike because their colors and shapes are different.

sort (to put into groups) I can sort these shapes by their color.

group (a set or a collection) These shapes are all alike in color. They make a group.

Doing Math

- Invite your child to help you sort the laundry. Ask your child to tell how the clothes in each pile are alike. (*They are all white, or all dark.*) Then ask what word(s) can be used to name each pile (*dark clothes, jeans*).

3

Playing Math

Here is a game that you and your child can play.

Cut small pieces of paper or cardboard, and use them to cover all the objects. Each player takes a turn picking up two covers. If the objects under the covers are alike, the player keeps the covers. If they are not alike, the player puts the covers back.

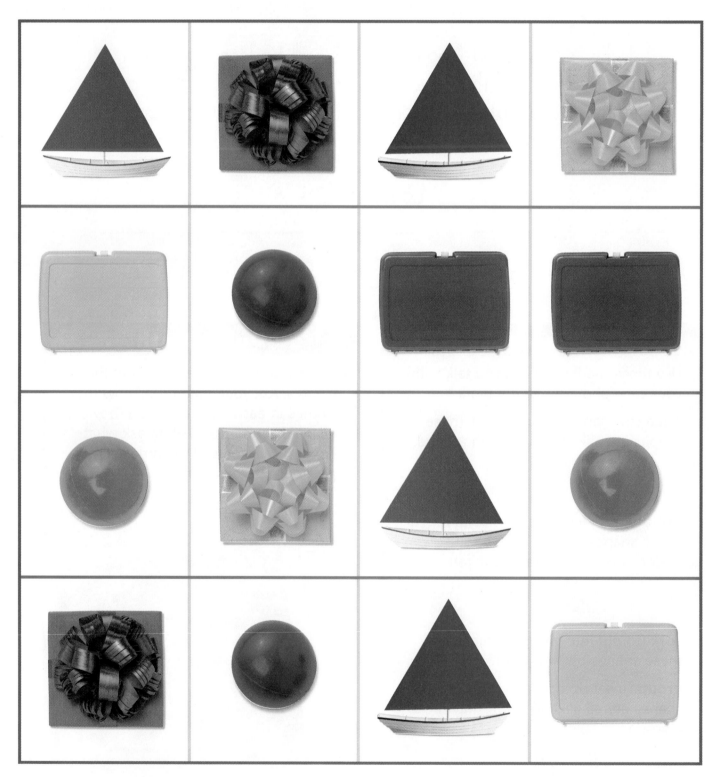

Name _____

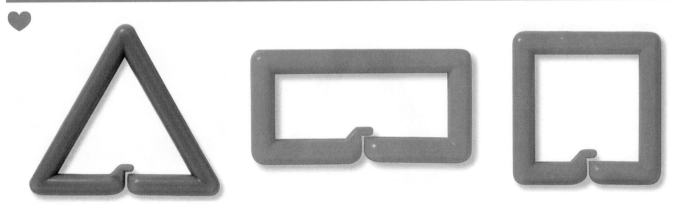

Circle the shapes that are alike in some way.

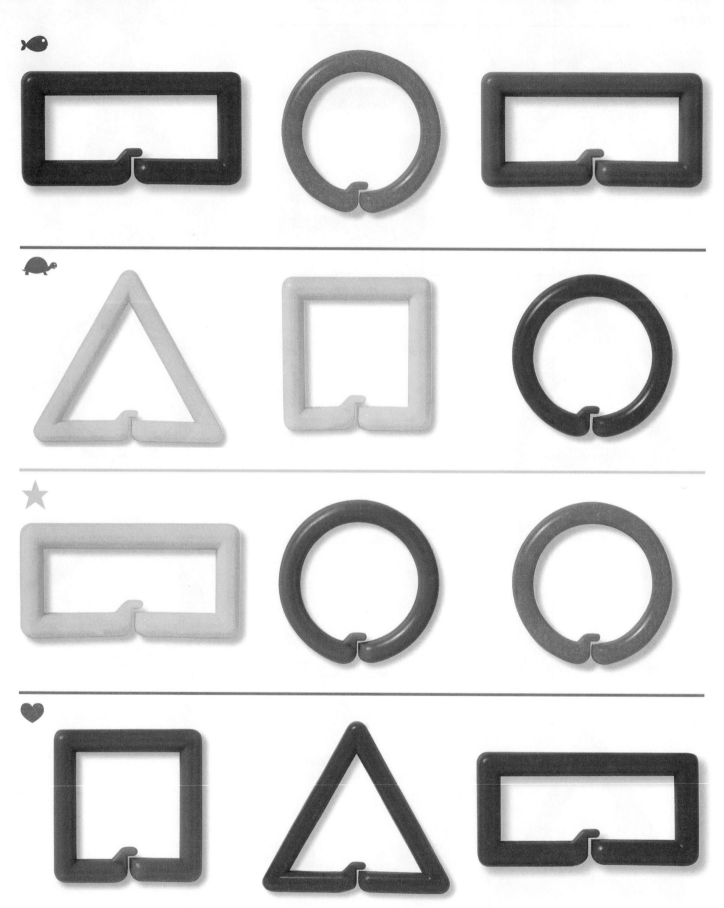

 Circle the shapes that are alike in some way.

Harcourt Brace School Publishers

HOME NOTE
Show your child two objects. Ask your child to tell you some ways in which they are alike.

6

More Sorting by Shape

Circle

Square

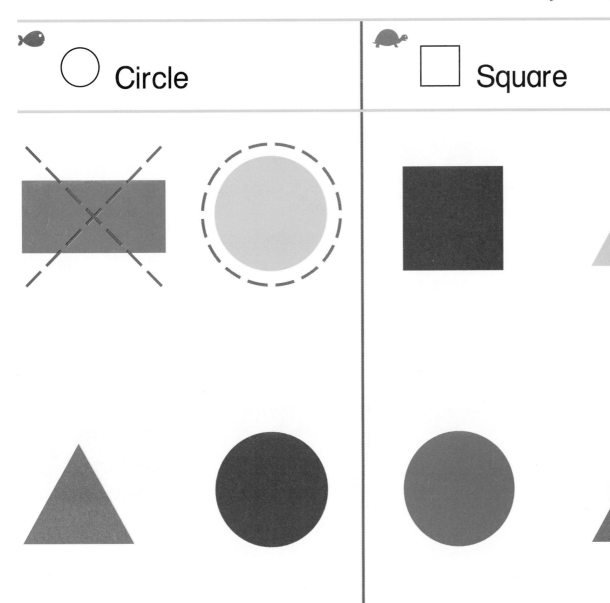

Circle the shapes that belong in the group. Mark an X on the shapes that do not belong.

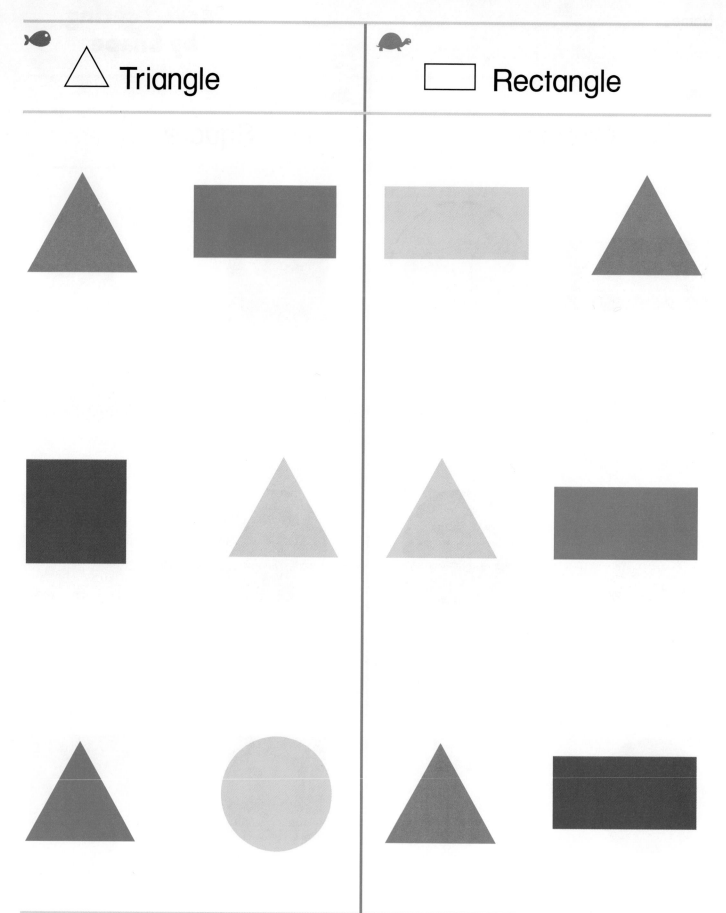

△ Triangle

▭ Rectangle

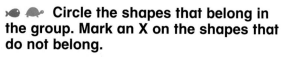 Circle the shapes that belong in the group. Mark an X on the shapes that do not belong.

8

 HOME NOTE
Invite your child to tell you how the shapes in each box are the same.

Harcourt Brace School Publishers

Sorting by Color and Shape

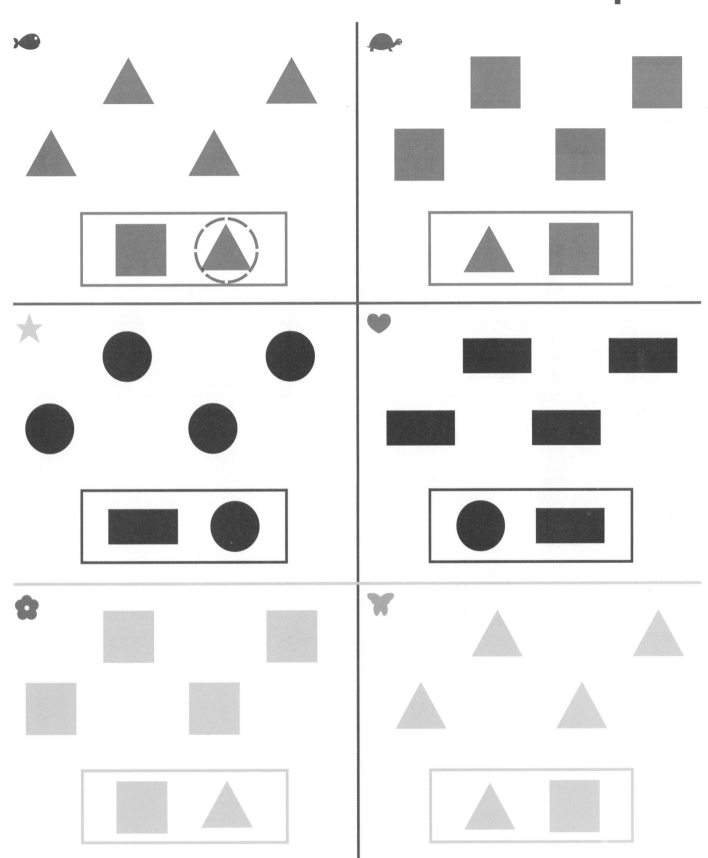

🐟 🐢 ⭐ ❤️ 🌸 🦋 **Circle the shape that belongs in the group.**

10

Harcourt Brace School Publishers

Sorting Everyday Objects

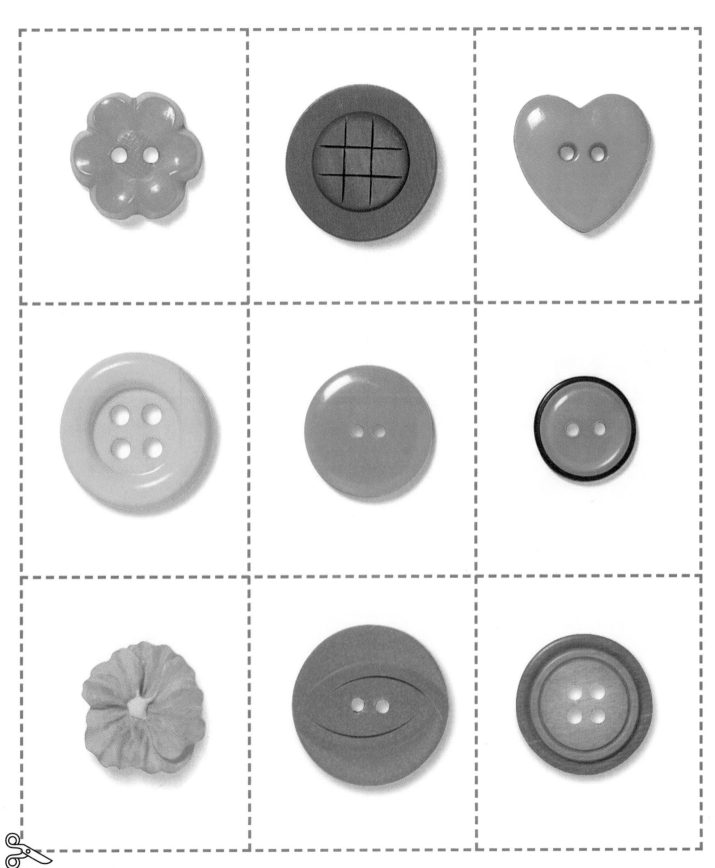

Cut out the cards and sort them.
Tell a partner how the buttons in each of your
groups are alike.

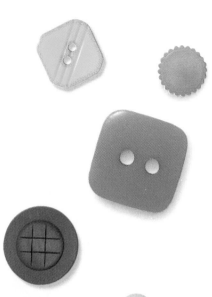

Things
That Are
Alike

By _____

HOME NOTE
Invite your child to share this book with you.
Ask your child the question on each page.

1

Which shapes are alike?

3

TAKE-HOME BOOK
Help children assemble the book.
Use the questions in the Teacher's Guide to have children
complete the book. Have children take the book home and share
it with family members.

CHAPTER ONE **13**

Which Attrilinks™ are alike in one way?

2

Harcourt Brace School Publishers

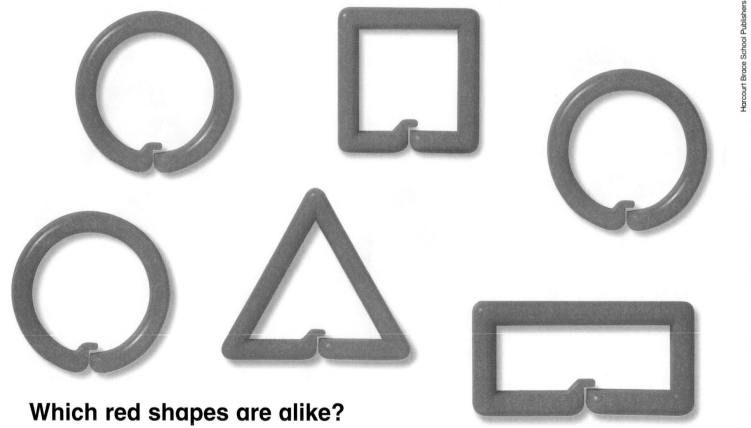

Which red shapes are alike?

4

14

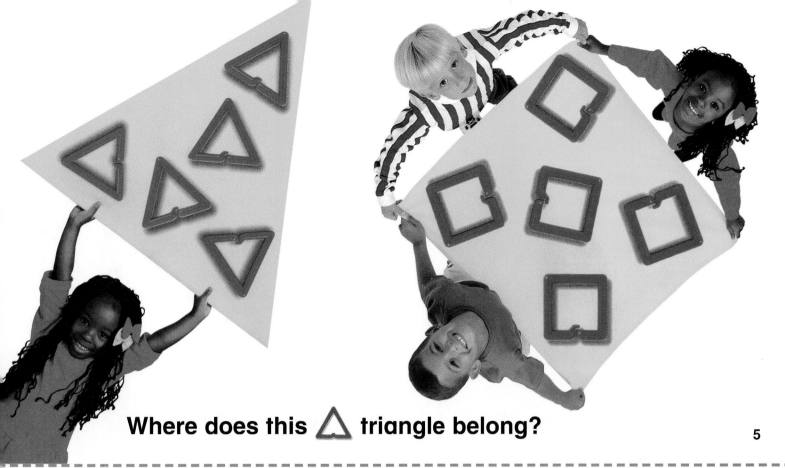

Where does this triangle belong?

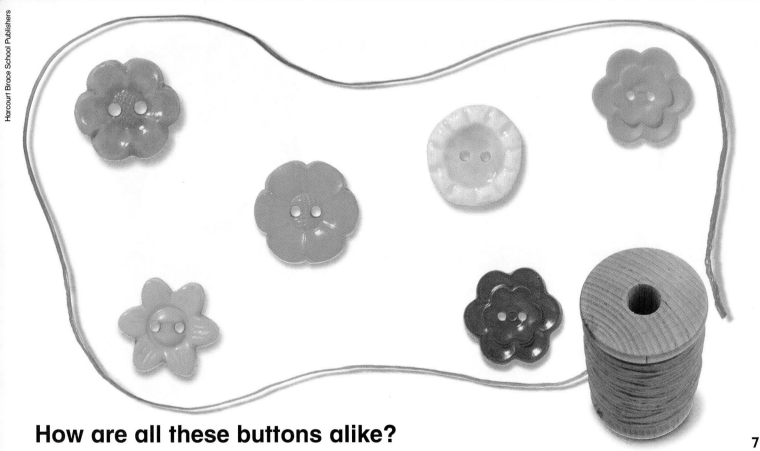

How are all these buttons alike?

How are the two buttons in each group alike?

I can draw two buttons that are alike.

MATH
ADVANTAGE

CHAPTER
2

Patterns and Movement

121212

🐟 **Color to copy the red/blue pattern.**
🐢 **Draw to copy the missing trees in the pattern.**

Photography Credits:

Harcourt Brace & Company Photographs

Key: (t) top, (b) bottom, (l) left, (r) right, (c) center.

3 (br) Greg Leary; 4 (tl) Terry Sinclair; 7, 8 Britt Runion; 9, 10, Rich Franco.

Illustration Credits:

Seth Larson: cover, 2; **Elizabeth Allen:** 9, 10; **Luisa D' Augusta:** 11, 12; **P.D. Cooper:** 5, 6; **Tony Griego:** 13, 16; **Steve Henry:** 3, 4.

Harcourt Brace School Publishers

Dear Family,
Today we started Chapter 2. We will copy and add to patterns and make up new ones. We will also make and tell about paths using the words *left* and *right*. Here and in the Home Notes are some things you can do to help me learn math.

Love,

_____

Talking Math

Use these math words as you talk with your child about his or her work.

pattern

path
This path turns to the right.

left   **right**

Doing Math

- Help your child get to know right from left by telling him or her which hand to use when handing you something. At the dinner table, for example, say "Pass me your plate with your left hand."

- When you sort laundry with your child, have him or her put dark clothes in a pile on the left side and light clothes in a pile on the right.

START

END

Playing Math

As a player moves along the path, he or she says the
pattern. When a player comes to a break in the path,
he or she must say what is missing. If correct, the
player moves on. If not, he or she starts again.

4

Circle the hat that comes next.

   Glue a picture to continue the pattern.

HOME NOTE
Invite your child to read or describe each pattern and tell which hat comes next.

Harcourt Brace School Publishers

Making Patterns

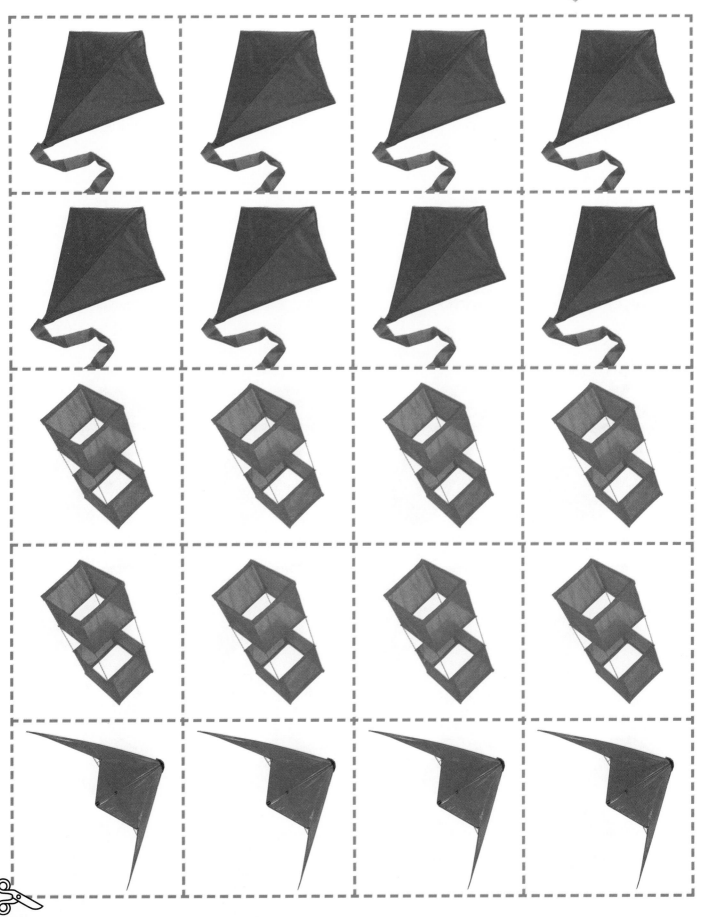

Cut out the pictures and sort them.
Make patterns.
Share your patterns with a classmate.

Harcourt Brace School Publishers

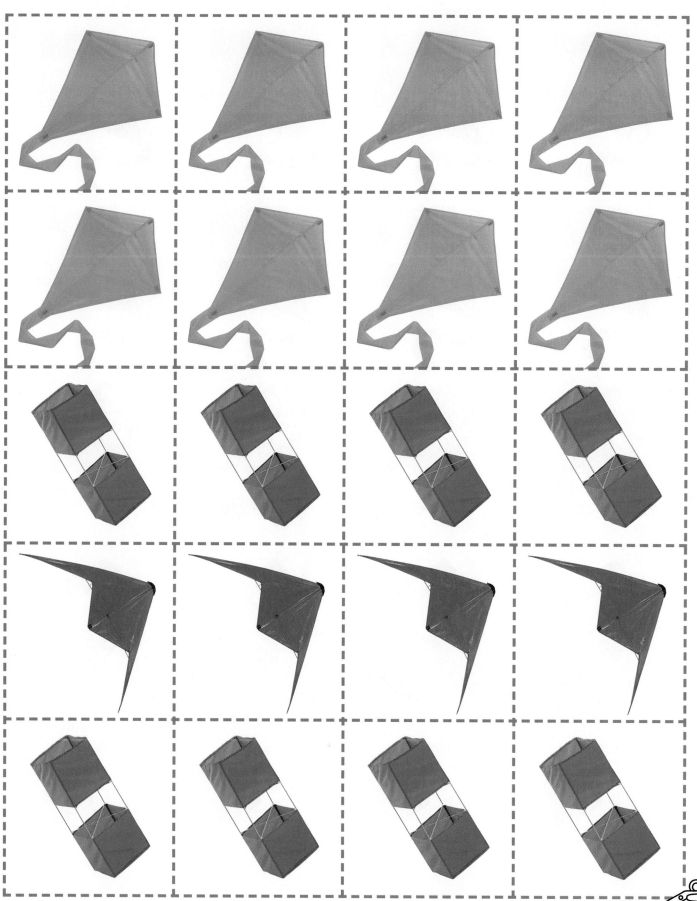

Cut out the pictures and sort them.
Make patterns.

8 Share your patterns with a classmate.

Harcourt Brace School Publishers

HOME NOTE
Ask your child to make some patterns.
Then have your child read or describe
each pattern to you.

Name _____

Left

Right

Put blue counters on the left. Put red counters
on the right. Draw something blue in the sandbox
on the left. Draw something red in the sandbox
on the right.

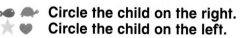 Circle the child on the right.
Circle the child on the left.

 HOME NOTE
Have your child describe the pictures,
using the words *left* and *right*.

10

Left and Right of a Path

Draw yourself and a friend to the left of the path.
Draw a house and a tree to the right of the path.

Cut out the pictures below.
Glue the pictures that are blue to the left of the road.
Glue the pictures that are red to the right of the road.

HOME NOTE
Take a walk outdoors with your child. Take turns pointing out things to the left and to the right of your path.

Harcourt Brace School Publishers

Paths and Patterns

By _____

HOME NOTE
Invite your child to share this book with
you. Read the words, and have your child
do what they say.

I can read this pattern.

3

✂ **TAKE-HOME BOOK**
Help children assemble the book.
Use the questions in the Teacher's Guide to have children
complete the book. Have children take the book home and share
it with family members.

I can copy this pattern.

Harcourt Brace School Publishers

I can show you which comes next.

14

I can draw a pattern.

5

I can turn to the right and to the left on a path.

7

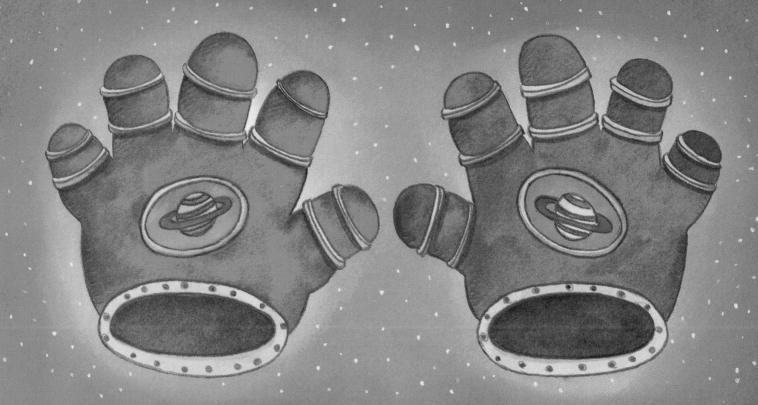

I can put my left hand here.

I can put my right hand here.

I can tell which side of the fence the dog is on.

MATH ADVANTAGE

CHAPTER
3
Matching and Counting

 Circle the two pictures that match.

Photography Credits:

Harcourt Brace & Company Photographs

Key: (t) top, (b) bottom, (l) left, (r) right, (c) center.

Cover, 2 Terry Sinclair.

Illustration Credits:

P.D. Cooper: 9, 10; **Liisa Chauncy Guida:** 3, 4; **Jennifer Beck Harris:** 5, 6; **Debra Melmon:** 7, 8; **Jim Paillot:** 13, 15.

Harcourt Brace School Publishers

Dear Family,
Today we started Chapter 3. We will learn how to match groups of objects to find out which groups have the same number, which have more, and which have fewer. We will learn to count objects, making sure we have counted each one. Here and in the Home Notes are things you can do to help me learn math.

Love,

_____

Talking Math

Use these math words as you talk with your child about his or her work in this chapter.

more than, fewer than
There are more birds than birdhouses.
There are fewer birdhouses than birds.

same as, how many?
There are the same number of bees as there are flowers.
How many bees do I see? I see three.

Doing Math

- Invite your child to count with you.

- As you set the table, have your child match things that will be used, such as forks and spoons, to see which groups have more, the same, or fewer.

- Have your child count the objects in a group. Watch to see that he or she counts one object for each number.

Playing Math

Start

Finish

DIRECTIONS: Count your way to the party. Have your child count the light rocks to get to each dark rock. If he or she misses a count, you take a turn. Then have your child count again.

4

Harcourt Brace School Publishers

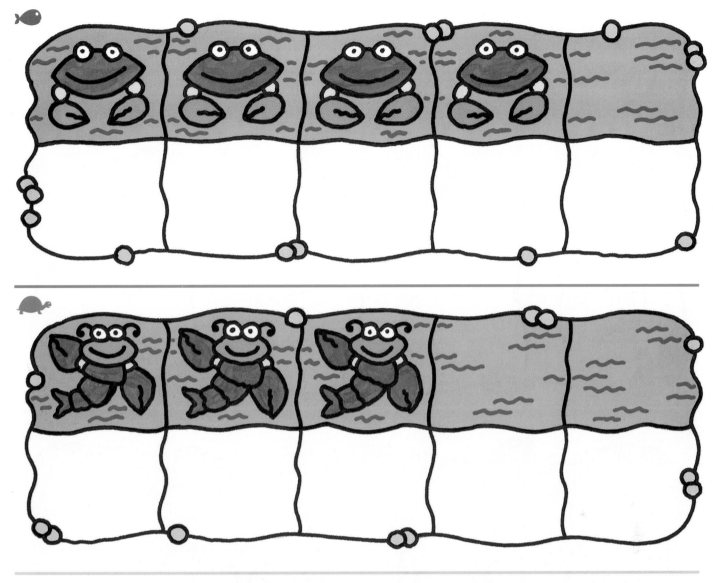

Place a counter on each picture.
Then move the counter to the box below it.
Color one box for each counter.

Draw a pot for each flower.
Draw an acorn for each squirrel.

Harcourt Brace School Publishers

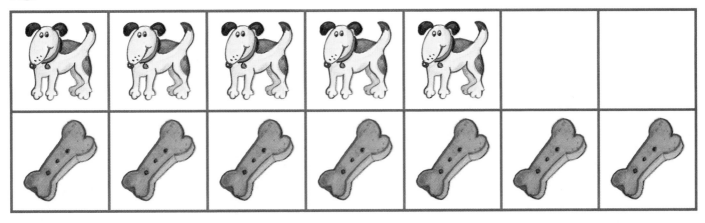

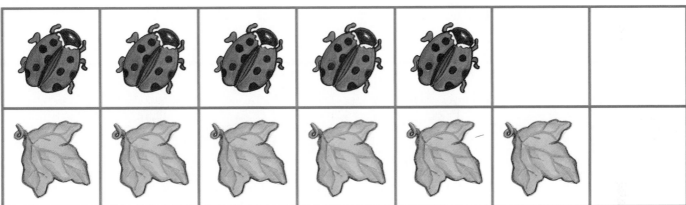

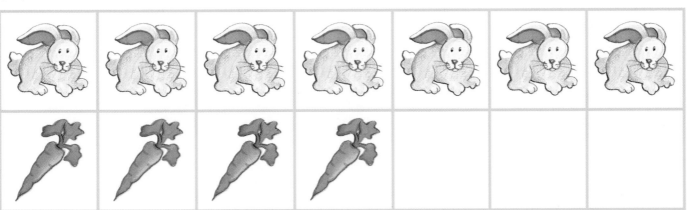

Put a counter on each picture.
Compare the number of counters in the two rows.
Circle the row that has more.

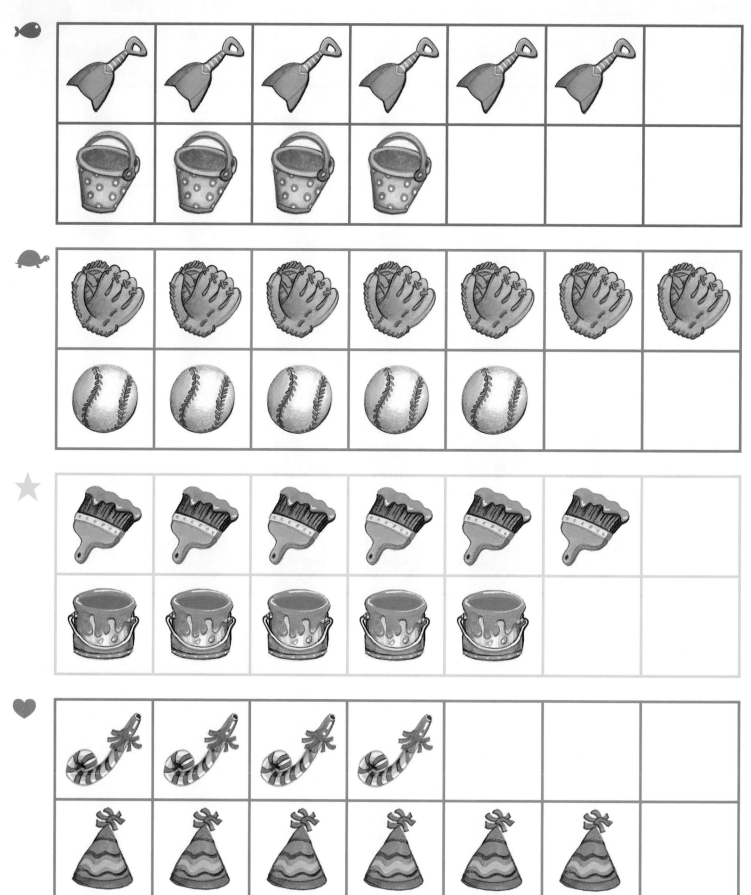

 Put a counter on each picture.
Compare the number of counters in the
two rows. Circle the row that has more.

HOME NOTE
Ask your child to match two groups
of objects to find out which group
has more.

Draw lines to match the animals in one group with the animals in the other group. Circle the group that has fewer.

 Draw lines to match the toys in one group with the toys in the other group. Circle the group that has fewer.

HOME NOTE
Have your child match two groups of objects to find out which group has fewer.

10

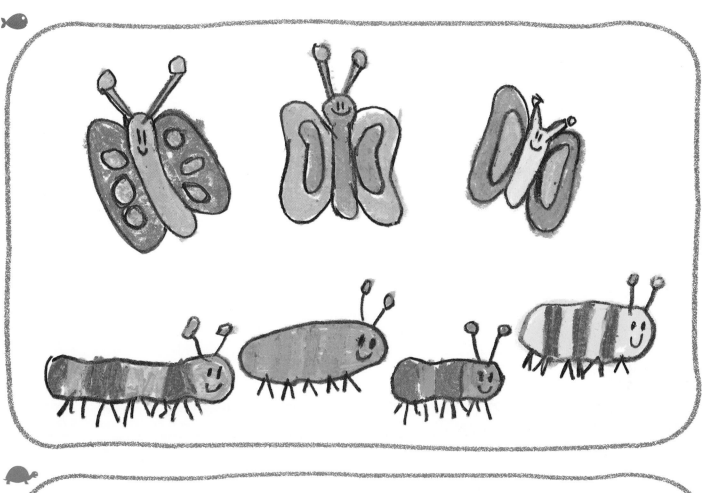

Count the pictures in each group. Circle the group that has one more.

 Count the objects in the group. Draw a group that has one more.

Harcourt Brace School Publishers

 **HOME NOTE**
Have your child count the objects in a group. Then add one, and have your child count again to find out how the group has changed.

I Can Count

By _____

HOME NOTE
Read the questions. Have your child count the objects and answer the questions.

1

Which group of fish has more?

3

TAKE-HOME BOOK
Help children assemble the book.
Use the questions in the Teacher's Guide to have children complete the book. Have children take the book home and share it with family members.

Is there a lily pad for each frog?

2

Which group has fewer?

4

How can you make a bee for each flower?

5

Which groups of butterflies have the same number?

7

How many birds?

6

Which group of flowers has more?

8

16

MATH
ADVANTAGE

CHAPTER
4

Numbers 0-5

Count the objects in each group.
Draw a red circle around the group that has more.
Draw a blue circle around the group that has fewer.

Photography Credits:

Harcourt Brace & Company Photographs

Key: (t) top, (b) bottom, (l) left, (r) right, (c) center.

3, 7 Victoria Bowen; 8 Sheri O'Neal.

Illustration Credits:

Tracy Sabin: Cover, 2; **Linda Davick:** 4; **Debra Spina Dixon:** 5, 6; **Dennis Greenlaw:** Cover title; **Nathan Jarvis:** 13, 16; **Scott Scheidly:** 11, 12; **Mary Thelen:** 9, 10.

Harcourt Brace School Publishers

Dear Family,
Today we started Chapter 4. We will learn about the numbers 0, 1, 2, 3, 4, and 5. We will match these numbers to groups of things. Then we will compare two groups to find out which number is greater and which number is less. Here and in the Home Notes are things you can do to help me learn math.

Love,

_____

Talking Math

You can help your child by using these words often in your daily activities.

greater than
4 is greater than 3.

4

less than
3 is less than 4.

3

Doing Math

- Have your child count small groups of objects such as shoes, cans, books, and chairs.

- Write the numbers 1, 2, 3, 4, and 5 on pieces of paper, and have your child match the numbers to the groups of things he or she has counted.

0 1 2 3 4 5

Playing Math

MATERIALS: pennies or bottle caps for markers

DIRECTIONS: Cut out the number cards 0–5, and mix them up.

4 Take a number card. Put a marker on each group the number stands for.

Name _____

 **1**  **2**  **3**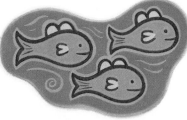

one   **two**   **three**

 1  **fish**

 2  **fish**

 3  **fish**

 2  **fish**

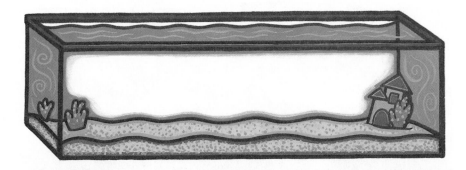

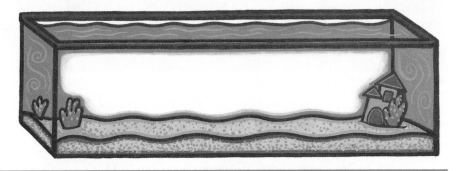

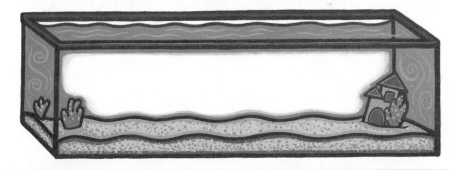

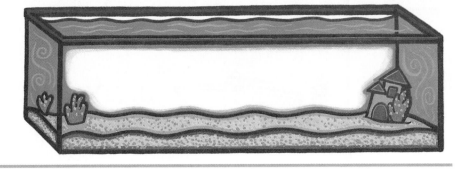

 Put a counter on each fish.
Read the number. Draw that many fish in the tank.

Harcourt Brace School Publishers

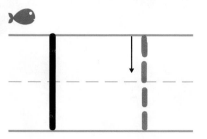  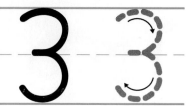

1 1   2 2   3 3

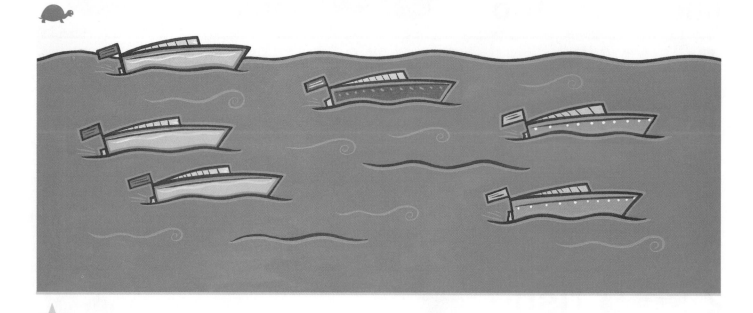

  yellow boats

  red boat

  orange boats

Trace the numbers.
Circle the boats to put them in color groups.
Write the number that tells how many boats are in each group.

6

HOME NOTE
Have your child find groups of 2 or 3 objects.

4 four

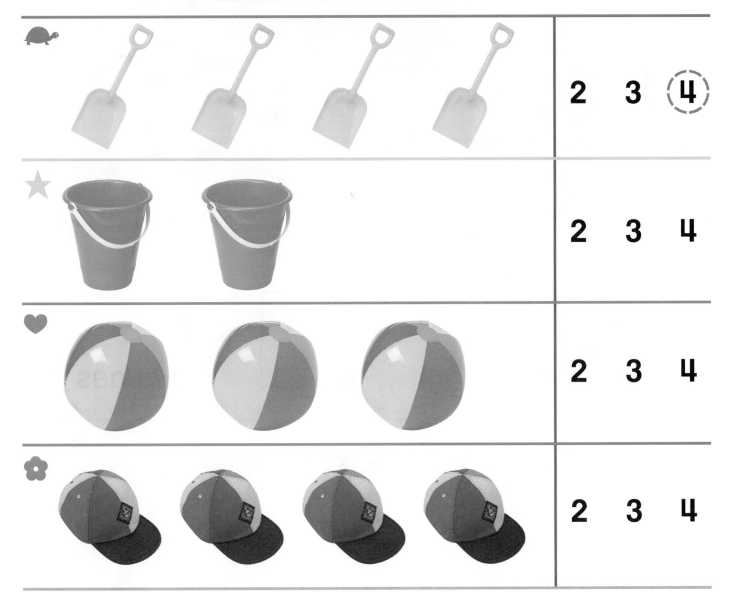

| | 2 | 3 | 4 |

Read the number. Put a counter on each child.

Circle the number that tells how many the children will take to the beach.

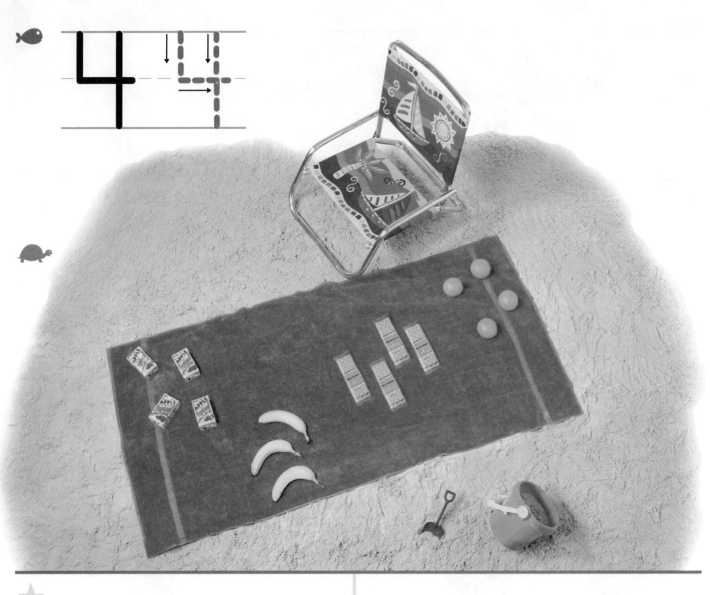

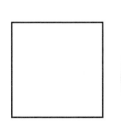 drinks

 oranges

 bananas

 crackers

🐟 Trace the number.
🐢 Circle each group of snacks.
⭐ Write the number that tells how many of each snack the children have.

 HOME NOTE
Have your child find groups of 4 objects.

8

Name _____

5 **five**

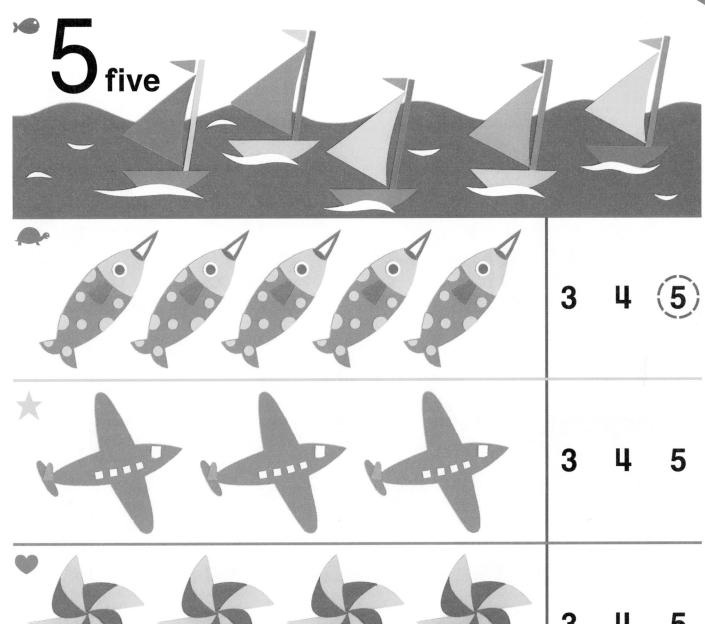

	3	4	⑤

| | 3 | 4 | 5 |

| | 3 | 4 | 5 |

| | 3 | 4 | 5 |

 Read the number. Put a counter on each sailboat.
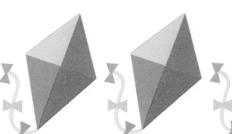 Circle the number that tells how many things are in the group.

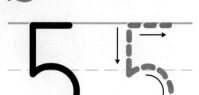

 green houses

 pink houses

Harcourt Brace School Publishers

 Trace the number.
Circle the group of green houses.
Circle the group of pink houses.

10 Write the number that tells how many houses there are of each color.

 HOME NOTE
Have your child find and label groups of 5 objects.

watermelons

oranges

peaches

bananas

apples

---

watermelon   apple

| 1 | (2) |

orange   peach

| | |

banana   orange

| | |

banana   peach

| | |

Write the number that tells how many fruits are in each group. Circle the number that is greater.

baskets   flowers

shells   flowers

hat   shells

birds   baskets

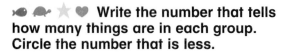 Write the number that tells how many things are in each group. Circle the number that is less.

 HOME NOTE
Have your child count two groups of objects, compare the numbers of objects in the two groups, and tell which number is greater.

Harcourt Brace School Publishers

5 Birds

By _____

Harcourt Brace School Publishers

HOME NOTE
Invite your child to share this number book.
Have him or her use the pictures for clues.

1

birds are splashing.

3

TAKE-HOME BOOK
Help children assemble the book.
Use the questions in the Teacher's Guide to have children
complete the book. Have children take the book home and share
it with family members.

CHAPTER FOUR 13

☐ birds are diving.

2

☐ birds are singing.

4

14

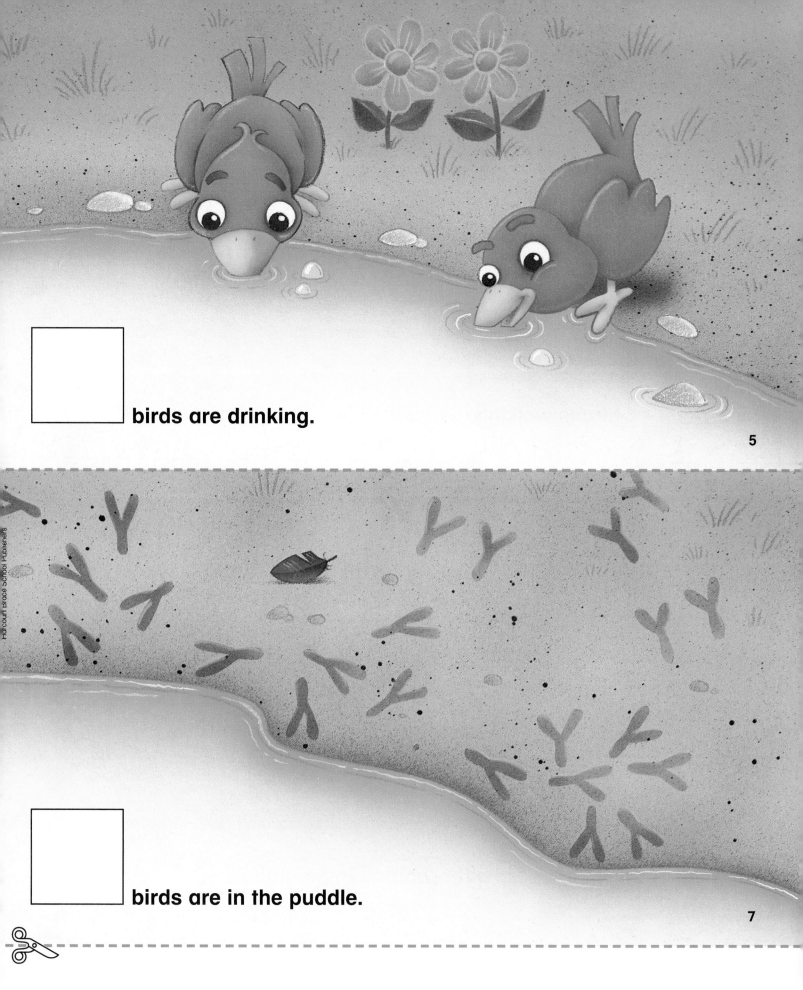

birds are drinking.

5

birds are in the puddle.

7

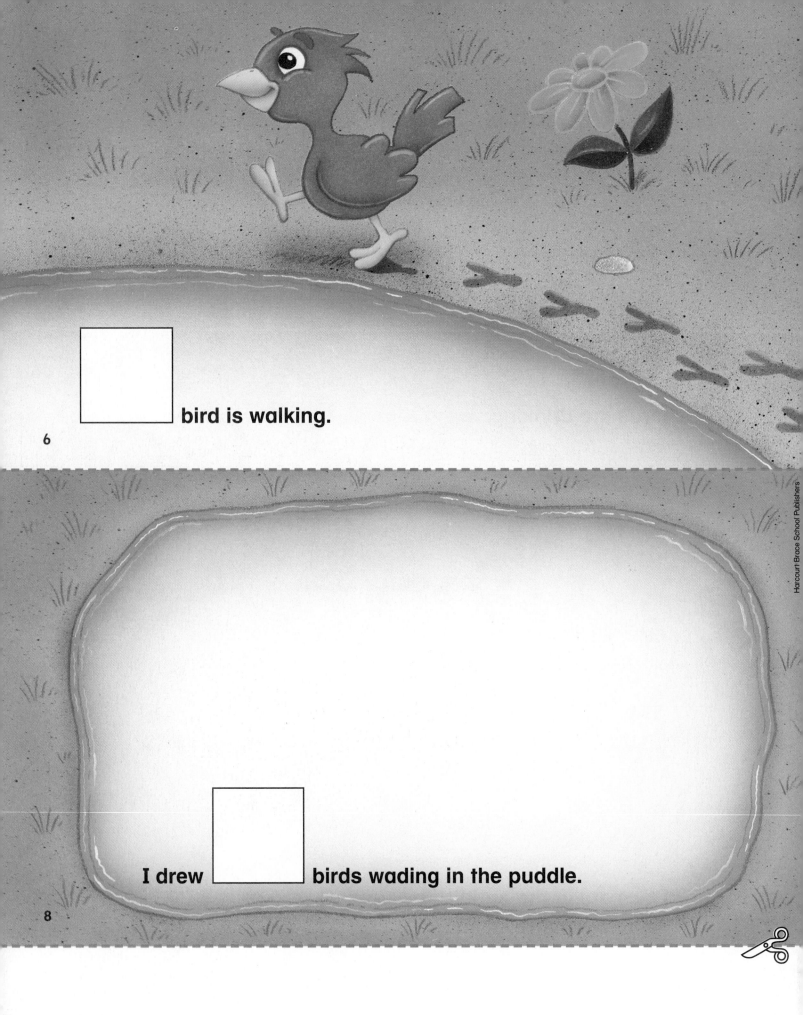

bird is walking.

6

I drew ⬜ birds wading in the puddle.

8

16

HARCOURT BRACE

MATH
ADVANTAGE

CHAPTER
5
Numbers
6-10

Circle the groups that have the same number of objects.

Photography Credits:

Harcourt Brace & Company Photographs

Key: (t) top, (b) bottom, (l) left, (r) right, (c) center.

3 Sheri O'Neal; 11, 12 (t) Bartlett Digital Photography; (bl), (br) Weronica Ankarorn; 13, 14 Sheri O' Neal.

Illustration Credits:

Kenneth Spengler: Cover, 2; **Tuko Fujisaki:** 7, 8; **Stacey Schuett:** 15, 16; **Mary Thelen:** 4, 6; **Lynn Titleman:** 9, 10.

2

Dear Family,

Today we started Chapter 5. We will learn numbers 6, 7, 8, 9, and 10. We will match these numbers to groups of things. Then we will compare the numbers of things in two groups to find out which number is greater and which number is less. Here and in the Home Notes are some things you can do to help me learn math.

Love,

_____

Talking Math

You can help your child by using these math words often in your daily activities.

greater than
8 is greater than 7.

less than
7 is less than 8.

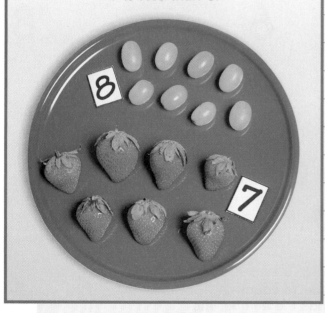

Doing Math

- Have your child count groups of six, seven, eight, nine, and ten objects.

- Write the numbers 6, 7, 8, 9, and 10 on pieces of paper, and have your child match the numbers to the groups of things he or she has counted.

- Have your child compare the objects in two groups. Ask your child which group has more.

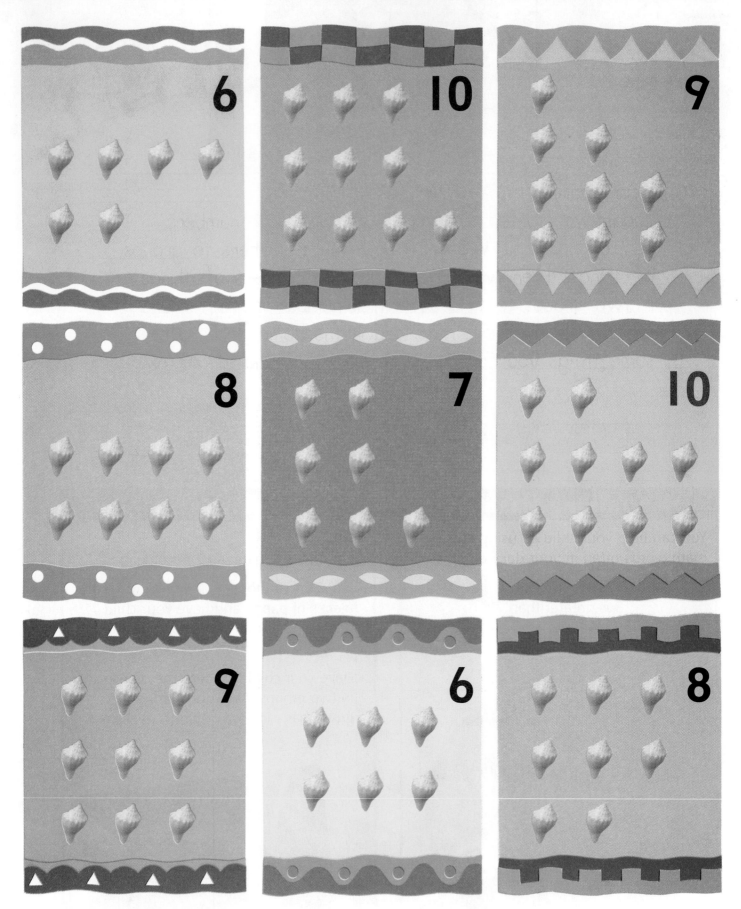

Playing Math

MATERIALS: coin to flip, bottle caps or other "shells" to count
One player flips the coin onto the gameboard, reads the number the
coin lands on, and counts out that many "shells." The other player
takes a turn, and then both players count their shells. The player
who collected more shells is the winner of that round.

4

Harcourt Brace School Publishers

Name _____

6 six

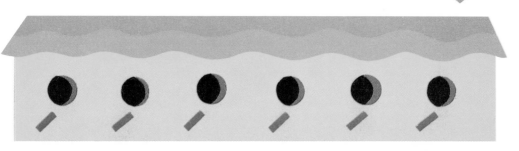

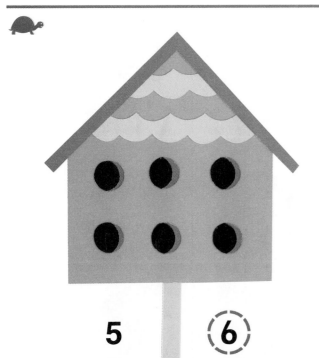

5   ⑥

5   6

5   6

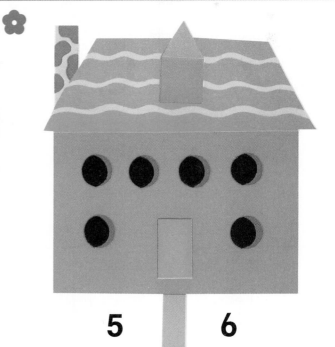

5   6

🐟 Read the number. Put a counter on each window.
🐢 ⭐ ❤ 🌸 Count the windows. Circle the number that
tells how many windows there are.

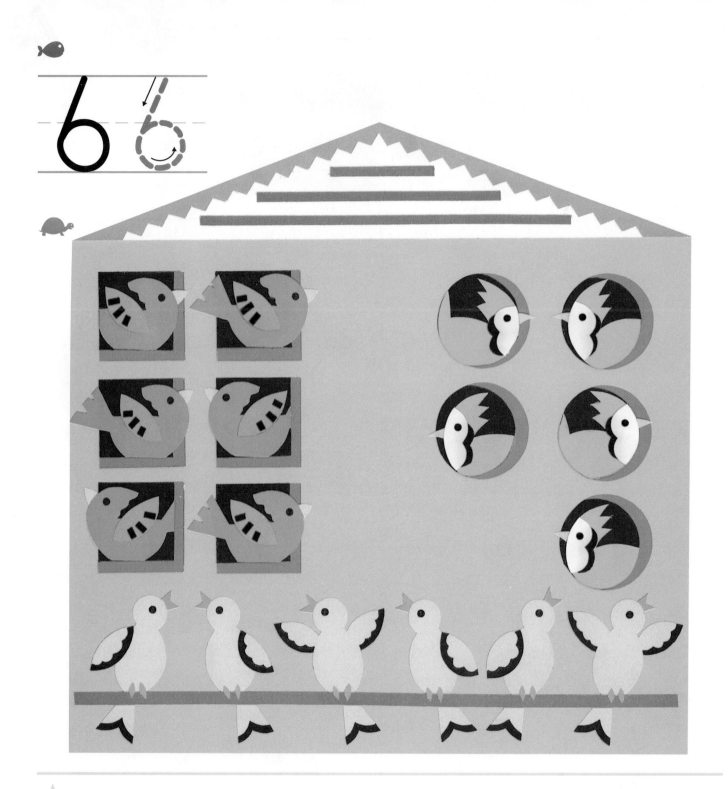

 birds

 birds

 birds

Trace the number.
Circle the groups of six birds.
Write the number that tells how many there are of each kind.

HOME NOTE
Have your child count out groups of 6 objects.

6

Harcourt Brace School Publishers

Name _____

7 seven

	⑥ 7
	6 7
	6 7
	6 7

Read the number. Put a counter on each bird.

Count the birds. Circle the number that tells how many birds there are.

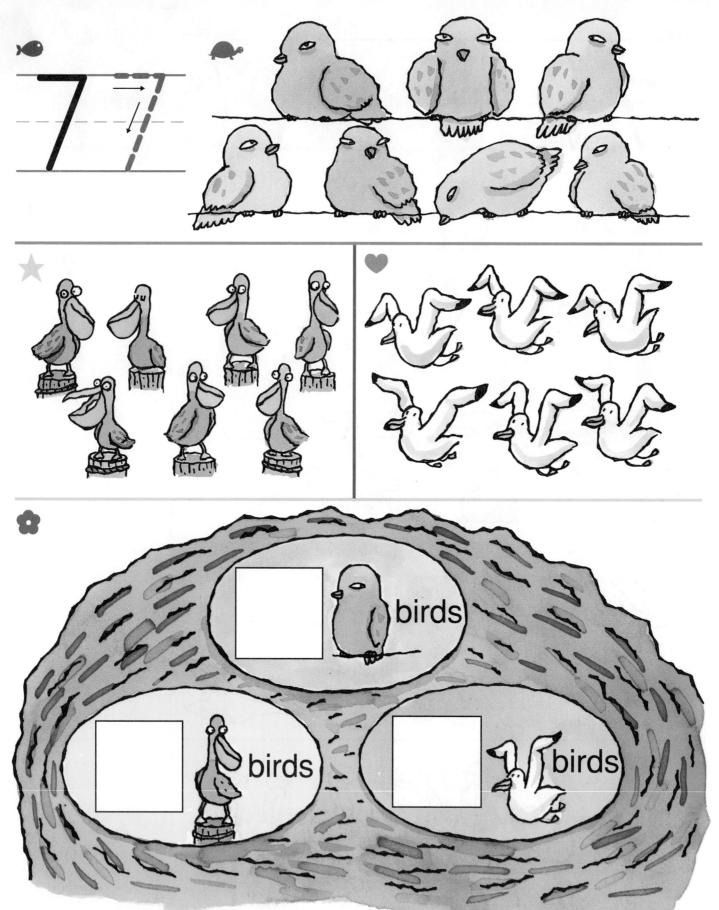

Trace the number.

Count the birds in the group.

Write the number that tells how many
there are of each kind.

8

Harcourt Brace School Publishers

 HOME NOTE
Have your child count out groups of
7 objects.

8 eight

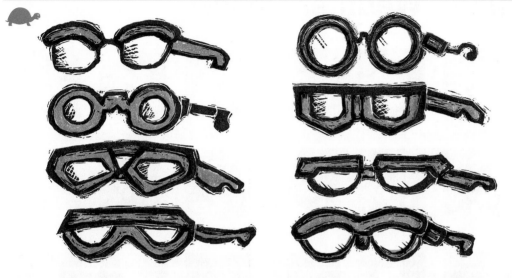

7 (8)

7   8

7   8

Read the number. Put a counter on each boot.

⭐ ❤ Count the objects. Circle the number that tells how many objects there are.

8

Island Weather Report

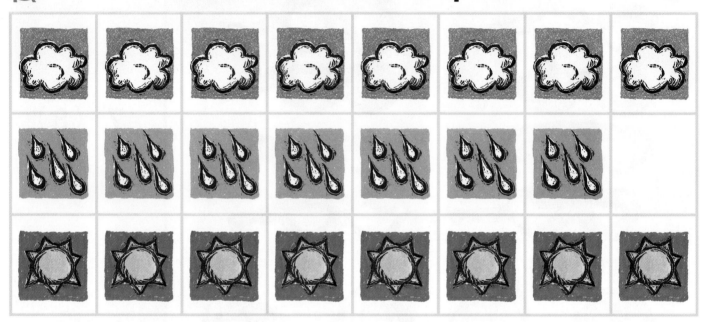

★  days were cloudy.

 days were rainy.

 days were sunny.

Harcourt Brace School Publishers

🐟 Trace the number.
🐢 Count the days of each kind in the graph.
★ Write the number that tells how many there were of each kind.

HOME NOTE
Have your child count out groups of 8 objects.

10

9 nine

⑦   8   9

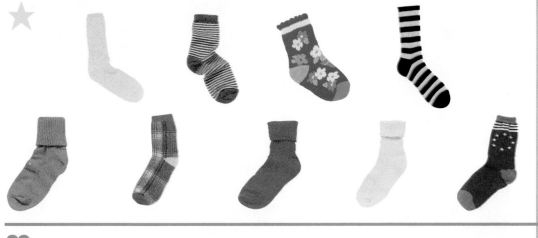

7   8   9

7   8   9

Read the number. Put a counter on each hat.

Count the pieces of clothing. Circle the number that tells how many there are.

Harcourt Brace School Publishers

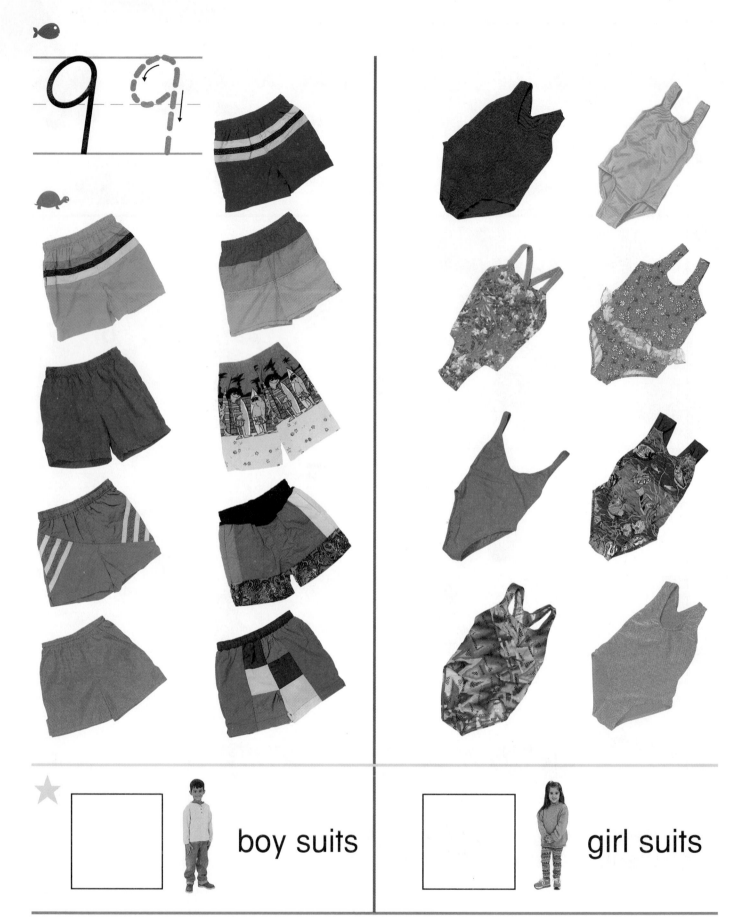

boy suits

girl suits

Trace the number.
Count the suits in each group.
Write the number that tells how many
there are of each kind.

HOME NOTE
Have your child count out groups of
9 objects.

10 **ten**

bears	9	10
dinosaurs	9	10
fish	9	10

Read the number. Put a counter on each bear.
Circle the groups of 10 toys.
Circle the number that tells how many there are of each kind.

★ I drew ☐ fish.

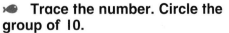 Trace the number. Circle the group of 10.
 Draw 10 fish.
★ Write the number that tells how many fish you drew.

HOME NOTE
Have your child count out groups of 10 objects.

Harcourt Brace School Publishers

I Saw 6 Fish Swimming

By _____

HOME NOTE
Have your child read the numbers.
Remind him or her to use the pictures
for clues.

children running,

3

TAKE-HOME BOOK
Help children assemble the book.
Use the questions in the Teacher's Guide to have children
complete the book. Have children take the book home and share
it with family members.

crabs crawling,

2

birds wading, and ☐ turtles sunning.

4

CHAPTER 6

Shapes
and Parts of Shapes

  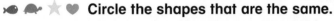 **Circle the shapes that are the same.**

Photography Credits:

Harcourt Brace & Company Photographs

Key: (t) top, (b) bottom, (l) left, (r) right, (c) center.

Cover Eric Camden; 3 (br), 4 Sheri O'Neal; 5 (lc) Don Couch; (rc) Victoria Bowen; All others Bartlett Digital Photography; 6 (tl) Sheri O' Neal; (bl) Victoria Bowen; All others Bartlett Digital Photography; 11 (inset) Sheri O' Neal; (c) Greg Leary; 12 (t), (bc), (b) Sheri O' Neal; (tc) Greg Leary; 13 top frame Victoria Bowen; (tl) Sheri O' Neal; (b) Bartlett Digital Photography; 14 (tl) Sheri O' Neal; (tr), (bcr), (br) Victoria Bowen; All others Bartlett Digital Photography; 15 Bartlett Digital Photography; 16 (lc), (rc), (b) Victoria Bowen; All others Bartlett Digital Photography.

Other: 15 (tc) Superstock

Illustration Credits: Julia Gorton: 7, 8;

Harcourt Brace School Publishers

Dear Family,
Today we started Chapter 6. We will learn to recognize and name some plane and solid shapes. We will learn about equal parts, too. Here and in the Home Notes are some things you can do to help me learn math.

Love,

_____

Talking Math

Use these math words as you talk with your child about his or her work.

Solid Shapes

cone

ball

can

box

Plane Shapes

circle

triangle

square

rectangle

Doing Math

- Choose one of the solid shapes listed, and help your child find objects that have that shape.

- Choose one of the plane shapes listed, and help your child find that shape on objects or in designs, such as wallpaper, curtains, and clothing.

Harcourt Brace School Publishers

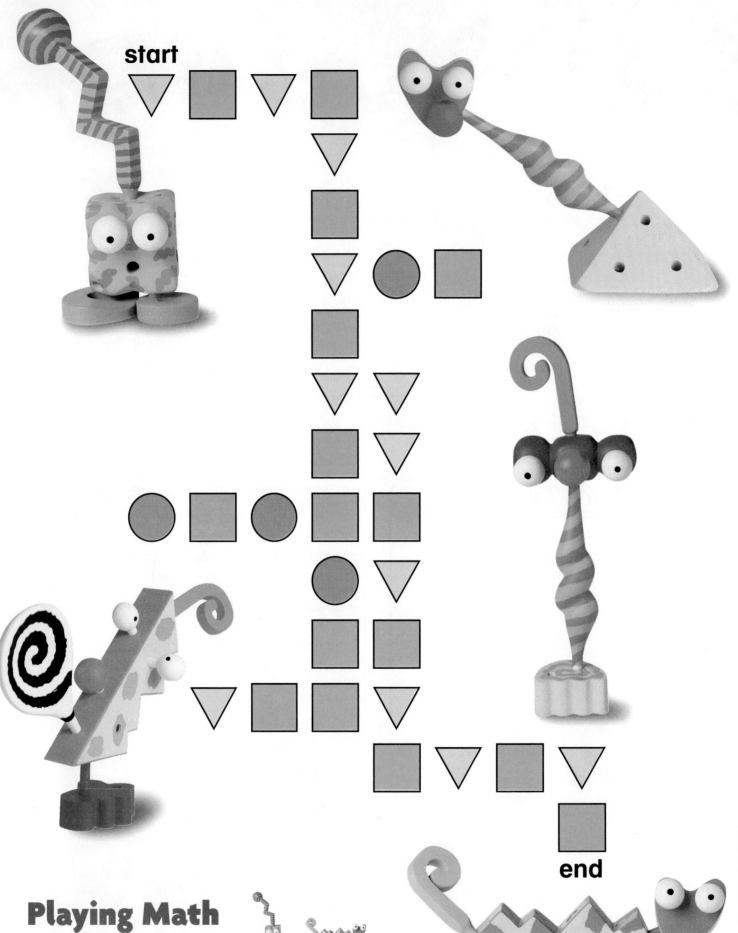

start

Playing Math

Draw a line to show the way from to .
Follow the pattern in the path.

end

4

ball **cone** **box** **can**

Use a blue crayon to circle the balls.
Use a red crayon to circle the cones.
Use a yellow crayon to circle the boxes.
Use a green crayon to circle the cans.

Look at the first shape. Circle the objects that are the same shape as the first object in the row.

6

HOME NOTE
Have your child tell you about the shapes of the objects using the words *ball*, *box*, *can*, and *cone*.

Sorting Solid Shapes

Cut out the cards.
Sort the shapes.

Cut out the cards.
Sort the shapes.

HOME NOTE
Have your child match the object
on a card to an object in the
house that is the same shape.

Harcourt Brace School Publishers

Color the triangles blue, the squares red,
and the rectangles yellow.

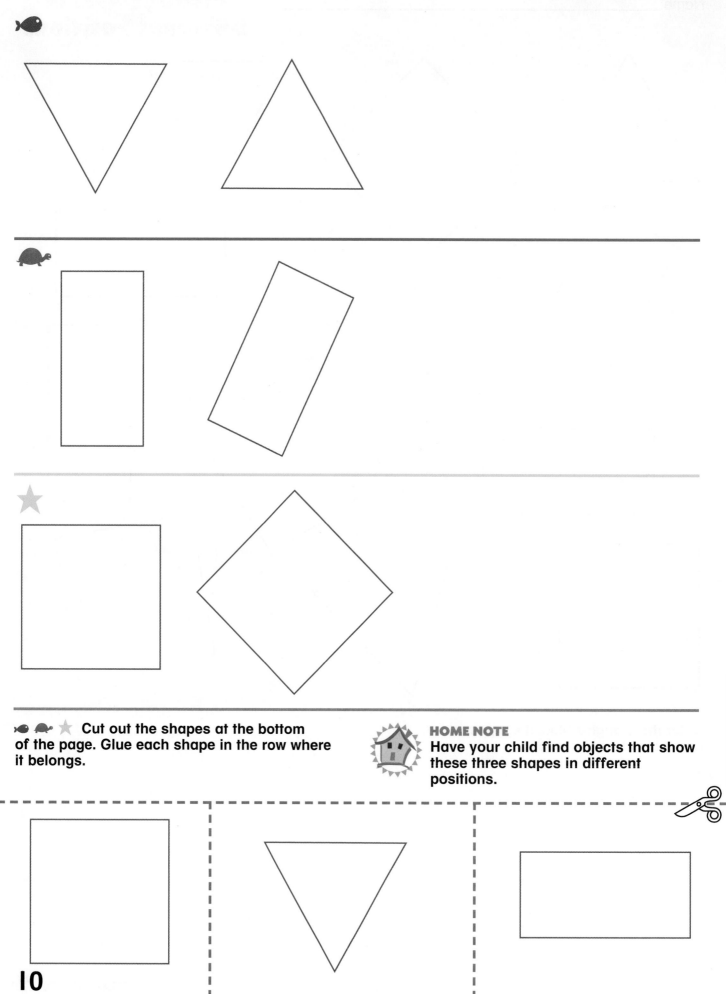

🐟 🐢 ⭐ Cut out the shapes at the bottom of the page. Glue each shape in the row where it belongs.

HOME NOTE
Have your child find objects that show these three shapes in different positions.

10

Equal and Not Equal Parts

Circle the cakes that show equal parts.

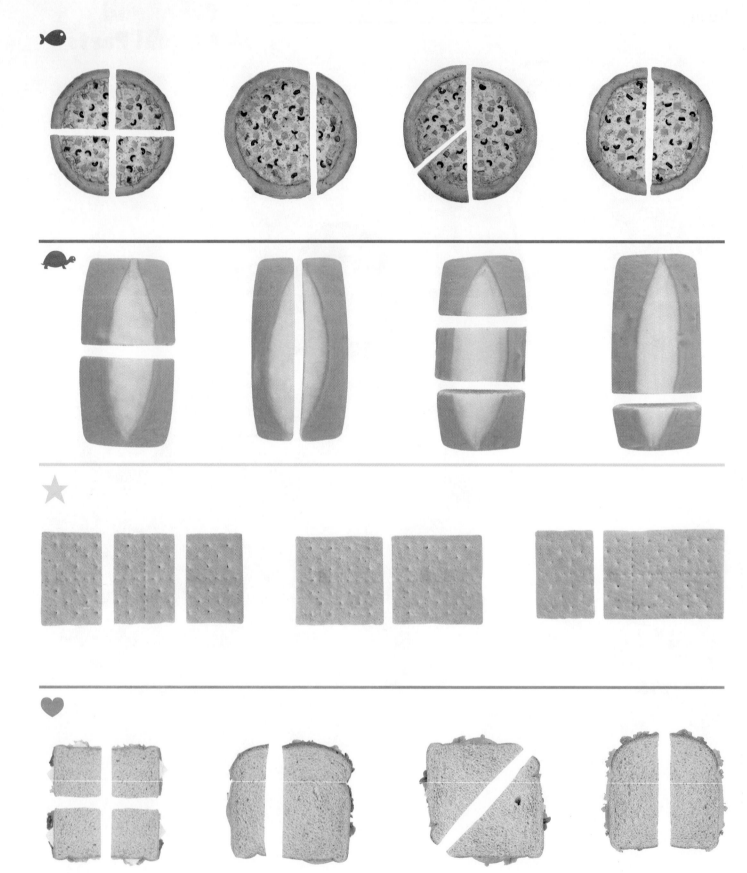

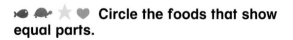 Circle the foods that show equal parts.

HOME NOTE
Have your child compare two foods that have been cut, one into equal parts and one into unequal parts, and talk about the difference.

Harcourt Brace School Publishers

Show Me Shapes

By _____

HOME NOTE
On each page, ask your child the questions,
and have him or her show you the shapes.

Harcourt Brace School Publishers

Which shapes are **cans?**

3

TAKE-HOME BOOK
Help children assemble the book.
Use the questions in the Teacher's Guide to have children
complete the book. Have children take the book home and share
it with family members.

Which shapes are 🔺 **cones?**

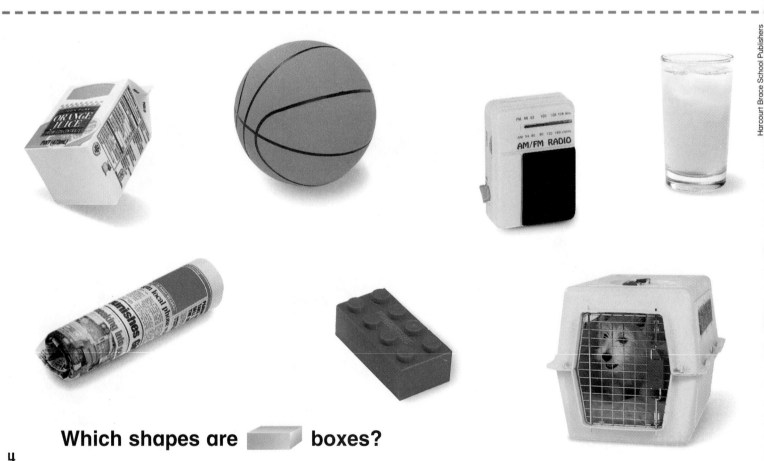

Which shapes are 📦 **boxes?**

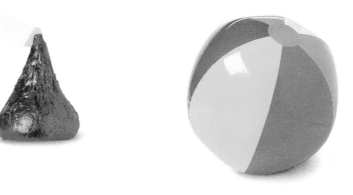

Which shapes are **balls?**

5

Harcourt Brace School Publishers

Which shapes are **triangles?**

7

Which shapes are ⬭ **circles?**

6

Which sandwich is cut into equal parts?

8

CHAPTER 7

Numbers 10-20

DIRECTIONS
Circle the groups of ten objects.

Photography Credits:

Harcourt Brace & Company Photographs

Key: (t) top, (b) bottom, (l) left, (r) right, (c) center.

Cover Eric Camden; 2 (tl) Eric Camden; (cl), (cr), (br) Don Couch; (tr) Bartlett Digital Photography; 3 (br) Don Couch; 5, 6, (b) Bartlett Digital Photography; (t), (c) Don Couch; 7 (t) Bartlett Digital Photography; (b) Don Couch; 8 (t), (b) Don Couch; (c) Weronica Ankaroron; 9 Bartlett Digital Photography; 11 Don Couch; 15, 16 Rich Franco.

Illustration Credits:

Dennis Greenlaw: Cover title, 13; **Steve Henry:** 7; **Jim Paillot:** 4; **Pam Perkins:** 9, 10.

Harcourt Brace School Publishers

Dear Family,
Today we started Chapter 7. In this chapter, we will count groups of 10. We will also learn to see numbers 11 to 20 as "10 and more." Here and in the Home Notes are some things you can do to help me learn math.

Love,

_____

Talking Math

Use these math words as you talk with your child about his or her work.

groups of 10

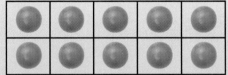

numbers 11 to 20

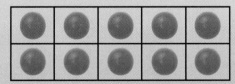

The names for 11, 12, 13, and 15 may be harder for children to remember since they do not contain the familiar number names *one, two, three,* and *five*.

Doing Math

- Count aloud often with your child.

- Have your child count out groups of ten objects, such as plastic straws, plastic chips, and pennies, and then count them by tens.

- Choose a number each day, and have your child do things during that day using that number. For example, have him or her form a group of nine toys, if 9 is the "number of the day."

START

3 6 3 2 5 7 4

4 5 6 7 3 4 6 8

2 5 5 8 6 4 3

END

Playing Math

MATERIALS penny, 20–30 bottle caps or other small objects

DIRECTIONS Players spin or flip the penny. If the penny lands heads up, the player moves 2 spaces; if the penny lands tails up, the player moves 3 spaces. Players read the number they land on and collect that many bottle caps. They "crawl down the tunnel" to get to the next row. At the end of the game, players count their bottle caps to see how many they have collected.

4

Harcourt Brace School Publishers

Name _____

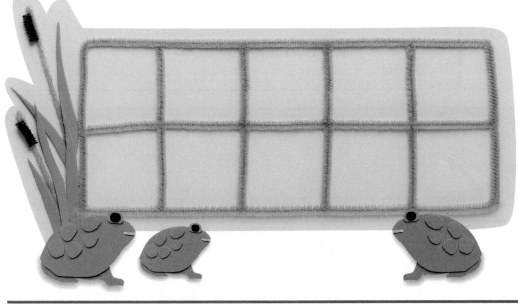

11   12

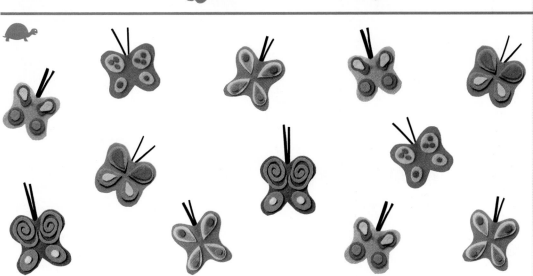

11   12

🐟 🐢 **Put a connecting cube on each object. Move the cubes to the 10-frame. Put a cube in each box and the extra cubes next to it. Count on from 10, and circle the number that tells how many.**

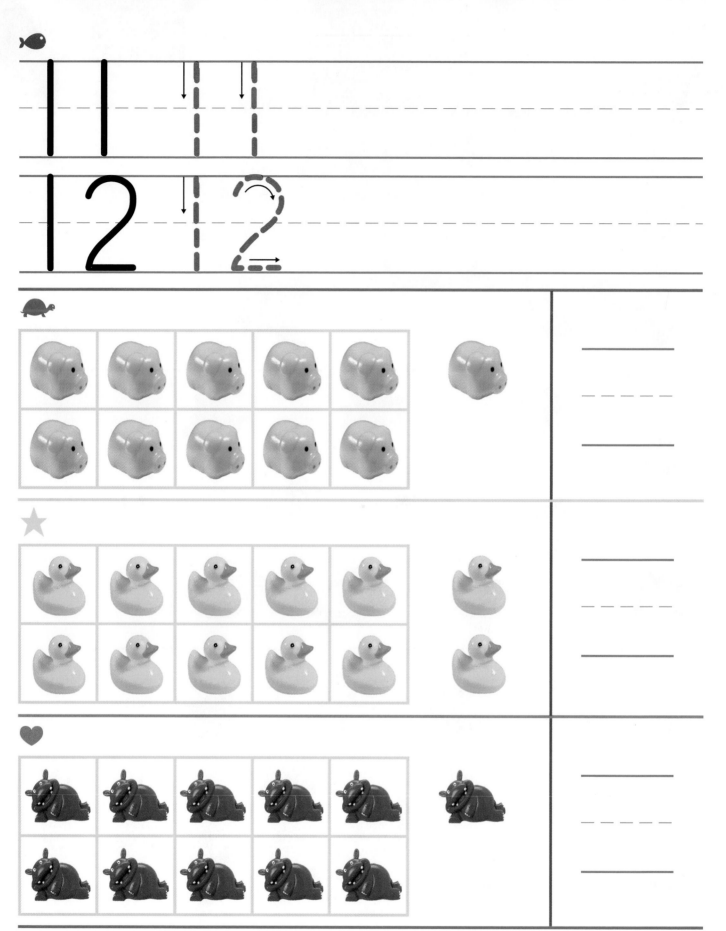

11

12

Trace and write the numbers.
Count the toy animals. Write the number that tells how many.

Harcourt Brace School Publishers

HOME NOTE
Have your child make a group of 11 objects by counting 10 and adding 1 more. Then have him or her count 10 and add 2 to make 12.

6

13   14

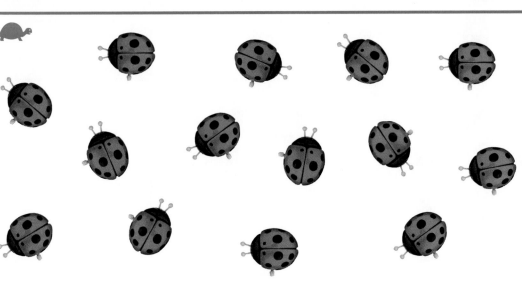

13   14

**Put a connecting cube on each toy.
Move the cubes to the 10-frame. Put one cube in
each box and the extra cubes next to it. Count on
from 10, and circle the number that tells how many.**

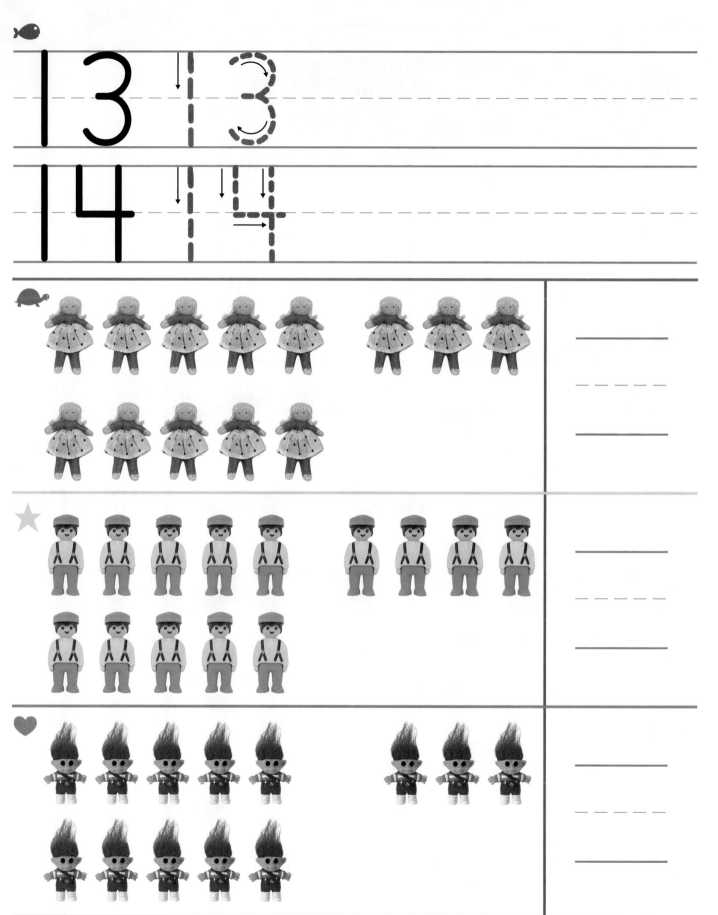

Trace and write the numbers.

Count the toys. Write the number that tells how many.

8

Harcourt Brace School Publishers

HOME NOTE
Have your child make a group of 13 objects by counting 10 and adding 3 more. Then have him or her count 10 and add 4 to make 14.

Name _____

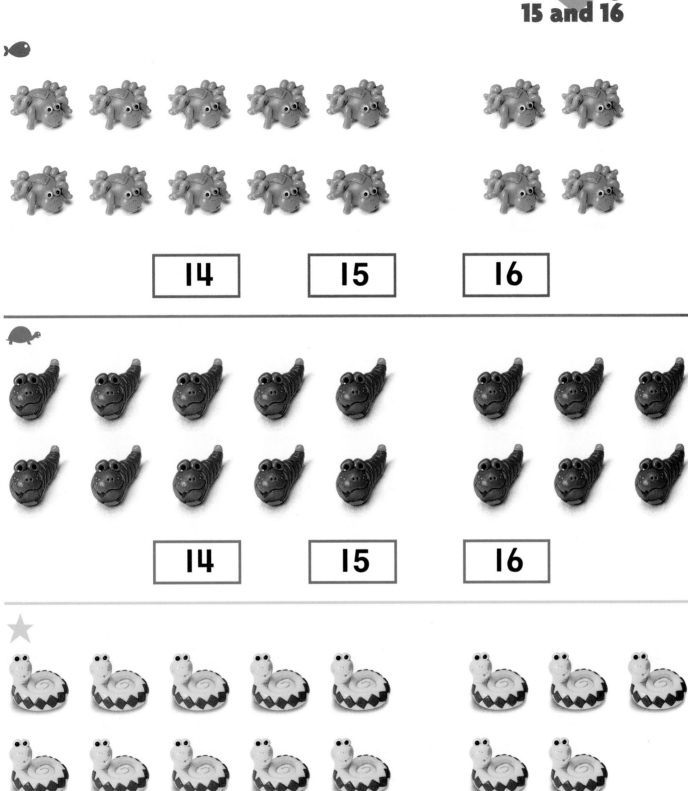

14   15   16

14   15   16

14   15   16

🐟 🐢 ⭐ Count the clay animals. Count on
from 10, and circle the number that tells
how many.

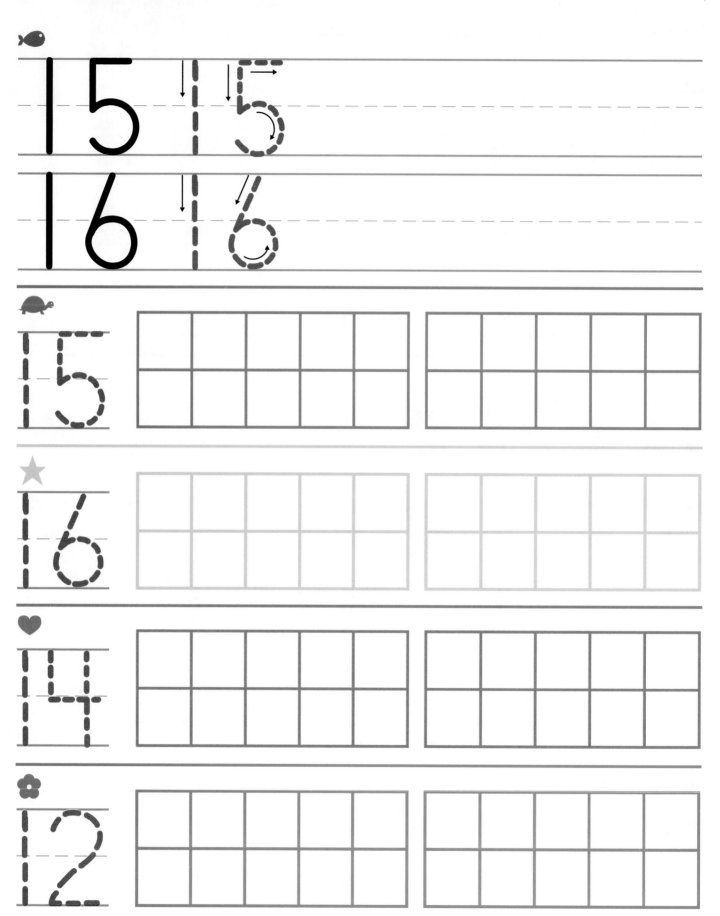

Trace and write the numbers.
Trace the numbers and color the squares to show how many.

HOME NOTE
Make a group of ten. Have your child add objects to make a group of 15 and then a group of 16.

Harcourt Brace School Publishers

16	17	18

16	17	18

16	17	18

Count the beads. Count on from 10, and circle the number that tells how many.

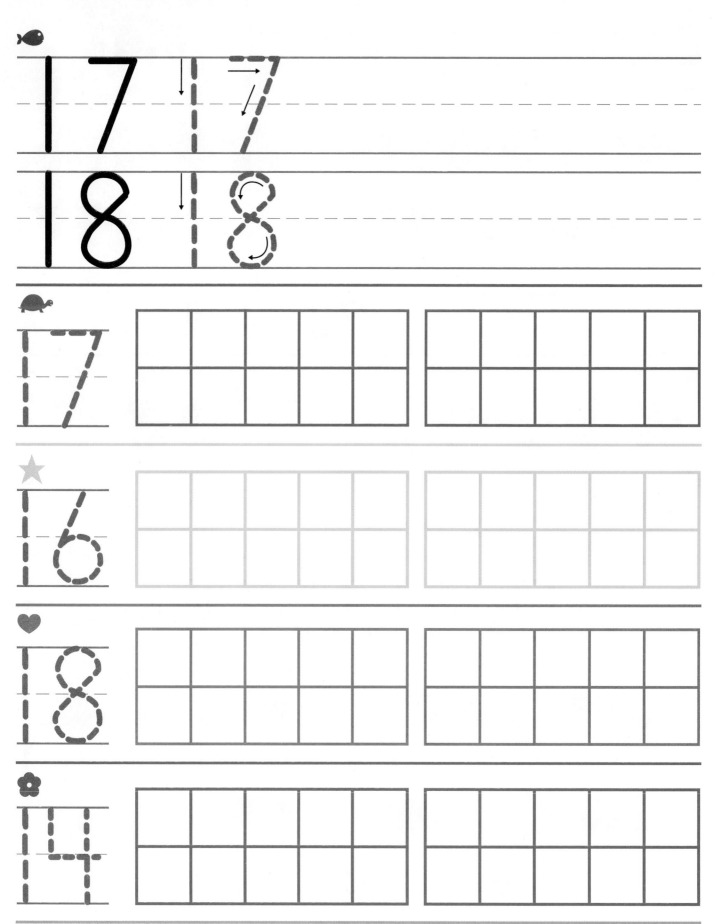

Trace and write the numbers.

Trace the numbers and color the squares to show how many.

Harcourt Brace School Publishers

HOME NOTE
Make a group of ten. Have your child add objects to make a group of 17 and then a group of 18.

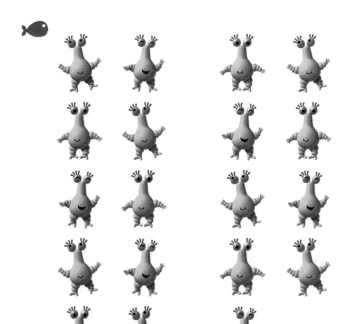

19	20

19	20

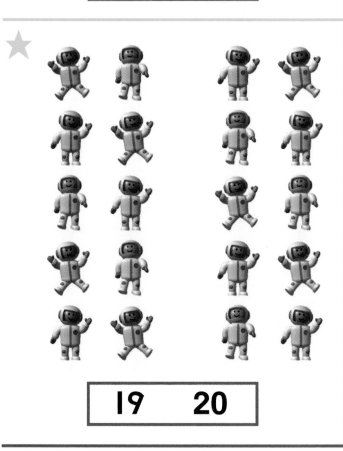

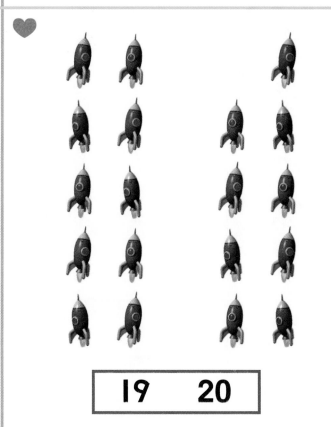

19	20

19	20

Count the space toys. Count on
from 10, and circle the number that tells
how many.

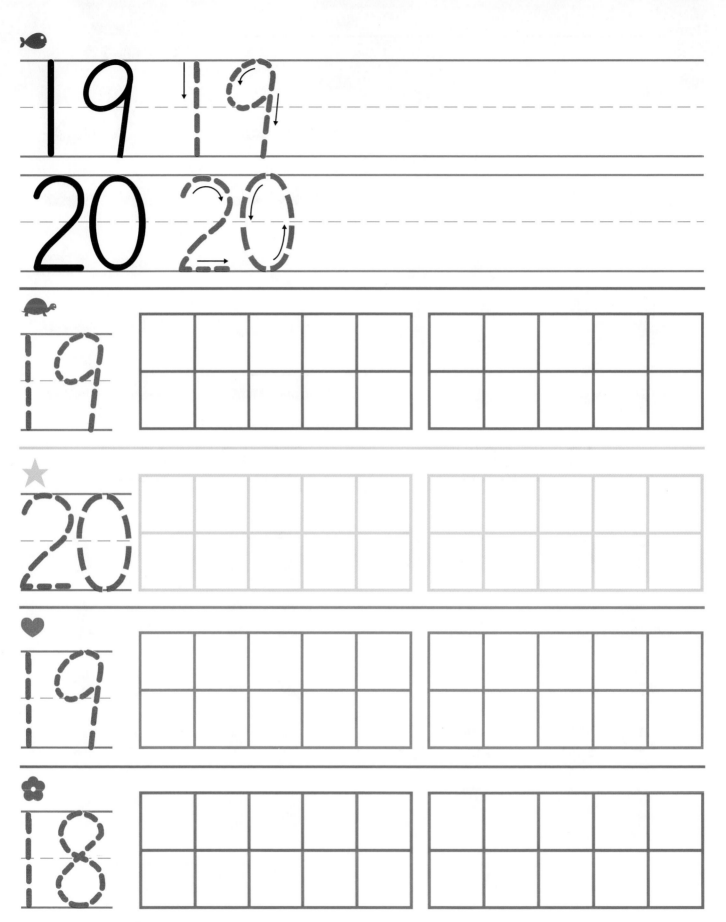

19

20

19

20

19

18

🐟 Trace and write the numbers.
🐢 ⭐ ♥ 🌸 Trace the numbers and color the squares to show how many.

HOME NOTE
Make a group of ten. Have your child add objects to make a group of 19 and then a group of 20.

14

Harcourt Brace School Publishers

How Many Fingers?

By _____

HOME NOTE
Read the "How Many?" question on each page for your child to answer.   I

How many prints?

_____

- - - - - - - - - - - - - - - -

_____

3

✂ **TAKE-HOME BOOK**
Help children assemble the book.
Use the questions in the Teacher's Guide to have children complete the book. Have children take the book home and share it with family members.

How many paints?

_____

- - - - -

_____

2

How many crayons?

_____

- - - - -

_____

4

16

MATH
ADVANTAGE

CHAPTER
8
Money

Apples
15¢

Cereal

ereal

  Circle the coins that are the same.

Photography Credits:

Harcourt Brace & Company Photographs

Key: (t) top, (b) bottom, (l) left, (r) right, (c) center.

Cover Terry Sinclair; 3 (br) Don Couch; 5, 7 Don Couch; 8 Victoria Bowen; 9 Bartlett Digital Photography; 10 Victoria Bowen; 11, 12 Don Couch.

Illustration Credits: Dennis Greenlaw: 13, 16; **Tracy Sabin:** 4.

Dear Family,
Today we started Chapter 8. We will learn about money—pennies, nickels, dimes, and quarters. We will pretend to buy things with these coins. Here and in the Home Notes are some ways you can help me learn math.

Love,

_____

Talking Math

You can help your child by using these math words often in your daily activities.

penny   **nickel**   **dime**   **quarter**

greater than
10¢ is greater than 5¢.

less than
5¢ is less than 10¢.

Doing Math

- Have your child sort a handful of change so that all the coins that are alike are together.

- Choose a different "coin of the day," and talk about what is shown on both sides of the coin.

- With your child, make up price tags of different amounts up to 10¢ for toys. Then give your child 10¢ in pennies to use to buy the toys. Later, add other coins.

Playing Math

MATERIALS: penny, nickel, dime, quarter; sock; game marker for each player
DIRECTIONS: Players take turns reaching into the sock, pulling out a coin, and moving their marker to the next picture of that coin. The first player to get to the bank is the winner.

¢
_____

¢
_____

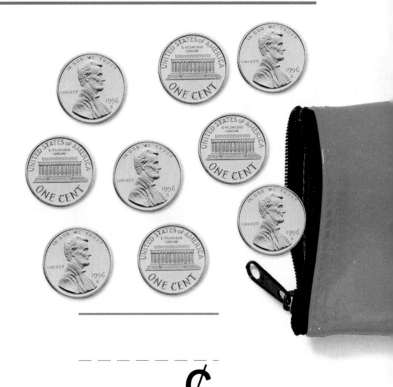

¢
_____

¢
_____

Count the pennies in each group.
Write the amount of money that is in each group.
Compare the amounts.
Circle the amount that is greater.

CHAPTER EIGHT **5**

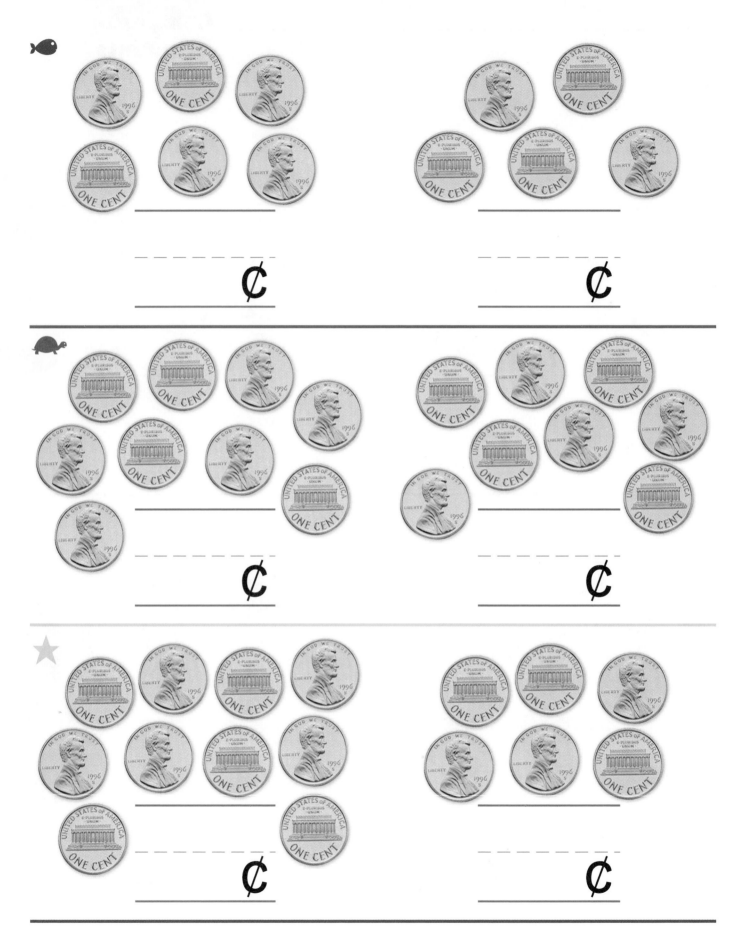

Count the pennies in each group. Write the amount of money that is in each group.
Compare the amounts.
Circle the amount that is less.

HOME NOTE
Have your child count two groups of pennies and then write the amount of money that is in each group. Ask him or her which amount is greater.

Harcourt Brace School Publishers

Penny

Nickel

Circle the nickels.

_____

_____ ¢

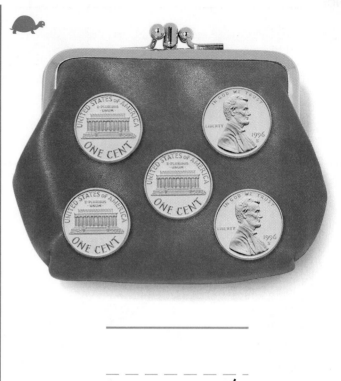

_____

_____ ¢

 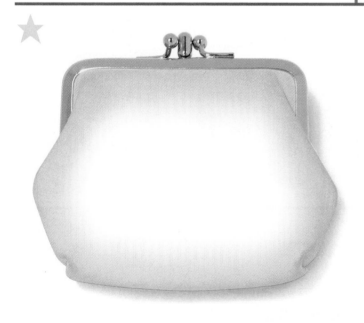

_____

_____ ¢

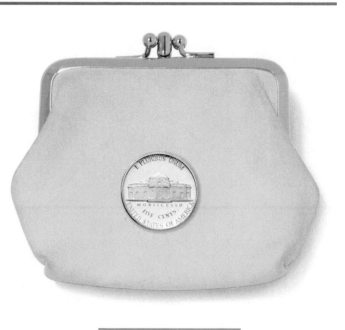

_____

_____ ¢

 Write how much money is in the purse.
Draw pennies in the empty purse to show
how many pennies are the same amount as a
nickel. Write how much money is in each purse.

 HOME NOTE
Ask your child to make a group of
pennies that is the same amount
of money as a nickel.

8

Harcourt Brace School Publishers

Dime

🐟 🐢 ⭐ 💛 🌸 🦋 **Circle the dimes.**

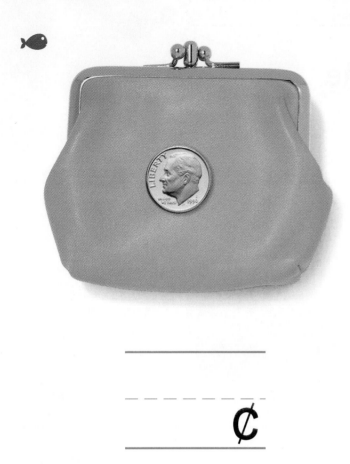

_____
- - - - - - - - -
¢

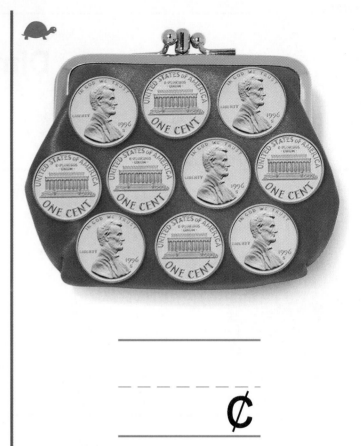

_____
- - - - - - - - -
¢

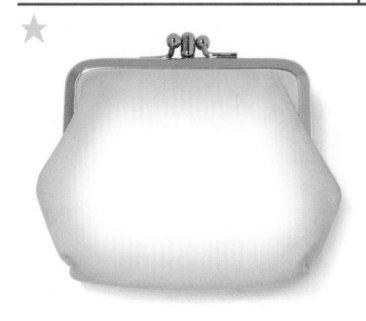

_____
- - - - - - - - -
¢

_____
- - - - - - - - -
¢

 Write how much money is in the purse.
 Draw pennies in the empty purse to show
how many pennies are the same amount as a
dime. Write how much money is in each purse.

10

 HOME NOTE
Ask your child to make a group of
pennies that is the same amount
of money as a dime.

Harcourt Brace School Publishers

Name _____

Quarter

Circle the quarters.

Harcourt Brace School Publishers

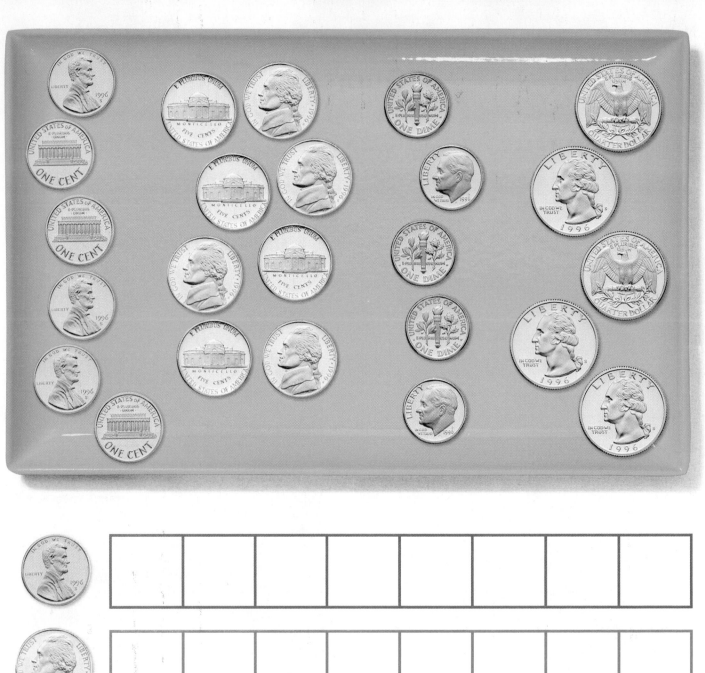

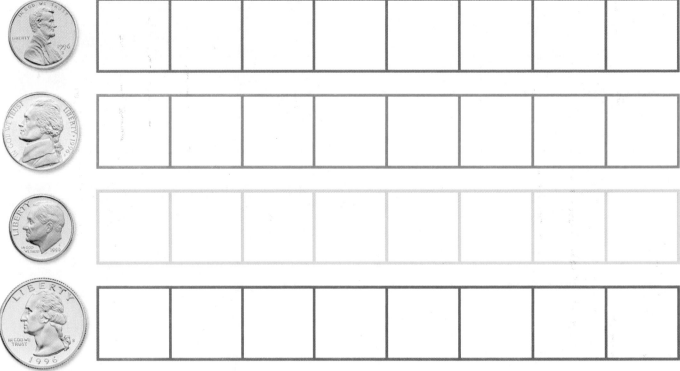

Count each group of coins.
Color squares next to each coin
to show how many there are of
that kind.

HOME NOTE
Have your child sort the change in
your pocket or purse by type of coin.

Harcourt Brace School Publishers

The Lost Coins

By _____

HOME NOTE

Invite your child to share this money book. Have him or her answer the question on each page.

The ladybugs asked, "How much is this?"

¢

3

TAKE-HOME BOOK
Help children assemble the book.
Use the questions in the Teacher's Guide to have children complete the book. Have children take the book home and share it with family members.

CHAPTER EIGHT 13

2 The spider asked, "Which coins are pennies?"

4 The fly asked, "Which coins are nickels?"

The bee asked, "How much is this?"

Harcourt Brace School Publishers

The butterfly asked, "How much is this?"

6 **The grasshopper asked, "Which coins are dimes?"**

8 **The ants asked, "Which coins are quarters?"**

Measuring

CHAPTER 9

MATH ADVANTAGE

Circle the objects that are as long as one paper clip.

Photography Credits:

Harcourt Brace & Company Photographs

Key: (t) top, (b) bottom, (l) left, (r) right, (c) center.

Cover Eric Camden; 2 ,3 (tl), 7, 8, 9, 10 Bartlett Digital Photography; 3 (bl) Greg Leavy; 3(br), 6 Victoria Bowen; 5 Sheri O'Neal.

Illustration Credits:

Dennis Greenlaw: 4; **Deborah Melmon:** 13, 16; **Pam Perkins:** 7, 8; **Don Sullivan:** 11, 12.

Harcourt Brace School Publishers

Dear Family,
Today we started Chapter 9. We will learn to measure how long and how tall things are. We will also compare how much different containers hold and how heavy different objects are. Here and in the Home Notes are some ways you can help me learn math.

Love,

_____

Talking Math

Use these math words as you talk with your child about his or her work in this chapter.

longer than, shorter than
The marker is longer than the crayon.
The marker is shorter than the pencil.

shortest, longest
The crayon is the shortest. The pencil is the longest.

shorter than, taller than
The cup is shorter than the glass. The glass is taller than the cup.

Doing Math

- Have your child compare the length of his or her hands and feet with those of other family members.

- Help your child compare an earlier height measurement with his or her height today.

- Have your child compare the amounts of water two plastic containers hold.

Playing Math

MATERIALS 2 kinds of small objects, such as bottle caps and coins as markers

DIRECTIONS Players guess whether an object on the board is longer or shorter than the game strip and then measure to check. If correct, the player places a marker on that square. The first player to fill three spaces in a row—across or down—is the winner.

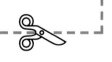

4

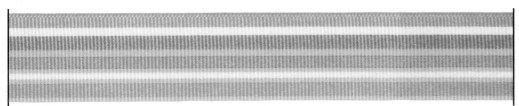

   ♥ Draw a line that is longer than the ribbon.

 Draw an object that is shorter than the one you see.

Harcourt Brace School Publishers

HOME NOTE
Have your child compare two objects and tell you which is longer and which is shorter.

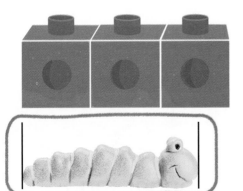

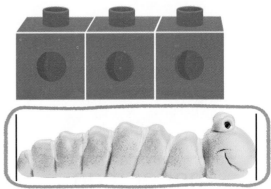

shorter than

longer than

🐟 🐢 ⭐ ❤ **Use 3 connecting cubes to measure each caterpillar. Circle with a blue crayon the caterpillars that are shorter than 3 cubes. Circle with a red crayon the caterpillars that are longer than 3 cubes.**

CHAPTER NINE **7**

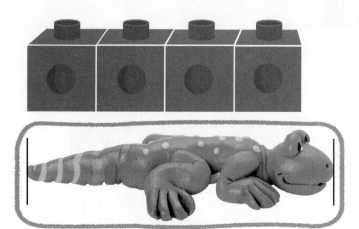

shorter than

longer than

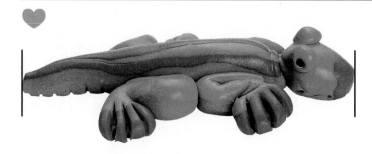 

Use 4 cubes to measure each lizard. Circle with a blue crayon the lizards that are shorter than 4 cubes. Circle with a red crayon the lizards that are longer than 4 cubes.

HOME NOTE

Give your child an object to use to measure other objects. Have him or her find objects that are longer than or shorter than the measuring object.

8

Harcourt Brace School Publishers

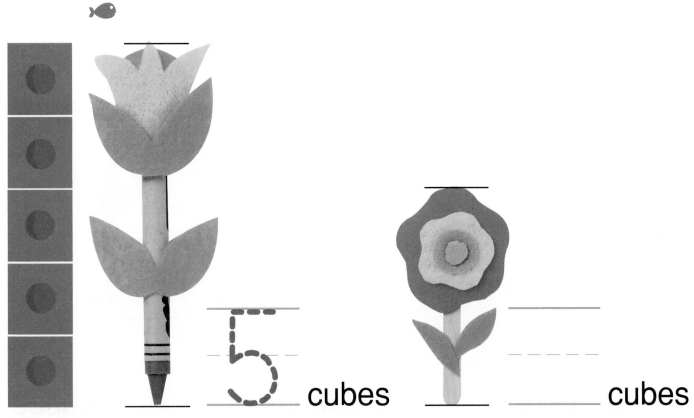

5 cubes

_____ cubes

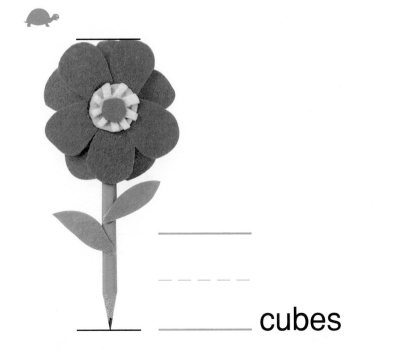

_____ cubes

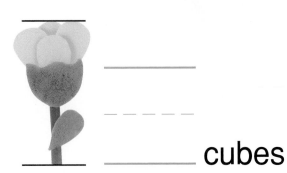

_____ cubes

Use connecting cubes to measure each flower.
Write the number of cubes tall. Circle the flower that
is taller.

_____ cubes _____ cubes _____ cubes

_____ cubes _____ cube _____ cubes

 Use connecting cubes to measure each flower. Write the number of cubes tall. Circle the flower that is the tallest.

10

Harcourt Brace School Publishers

About How Long?

Guess _____ cubes

Measure __4__ cubes

Guess _____ cubes

Measure _____ cubes

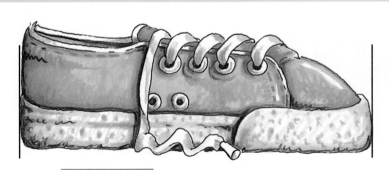

Guess _____ cubes

Measure _____ cubes

 About how many connecting cubes long is each shoe? Guess. Write the number on the red lines. Then use cubes to measure. Write the number on the blue lines.

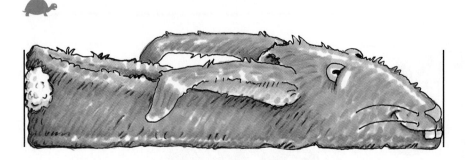

Guess _____ cubes   Measure _____ cubes

Guess _____ cubes   Measure _____ cubes

Guess _____ cubes   Measure _____ cubes

 About how many connecting cubes long is each slipper? Guess. Write the number on the red lines. Then use cubes to measure. Write the number on the blue lines.

12

HOME NOTE
Have your child guess how many paper clips long an object is. Then help him or her measure the object and compare the measurement with the guess.

Harcourt Brace School Publishers

The Three Bears' Measuring Book

By _____

1

Which chair is taller?

3

Which bear is shorter?

2

Which bear is the tallest?

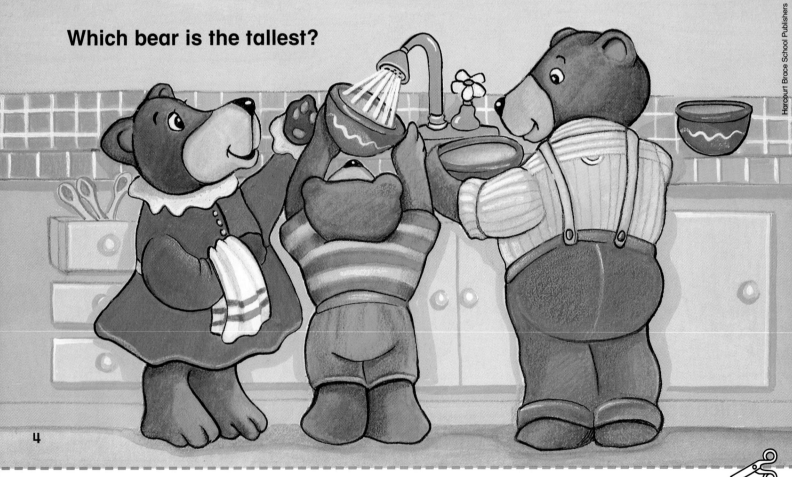

4

14

Which spoon is longer?

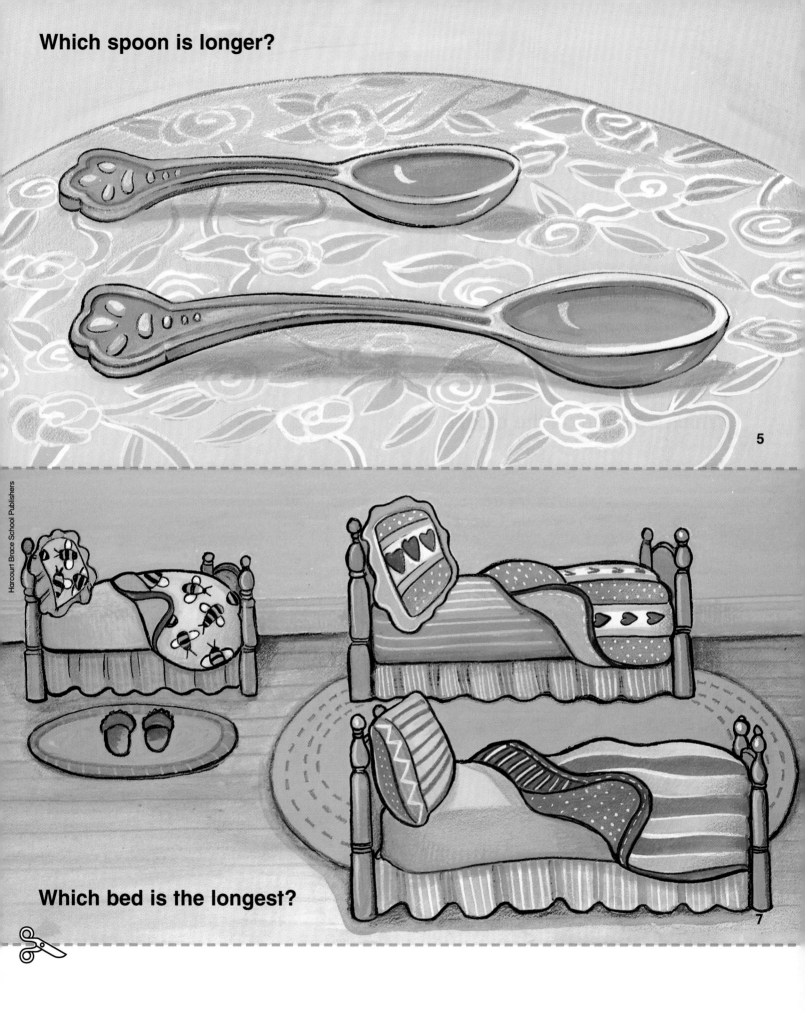

5

Which bed is the longest?

7

Harcourt Brace School Publishers

Which bowl holds the most?

6

Which bear is heavier?

8

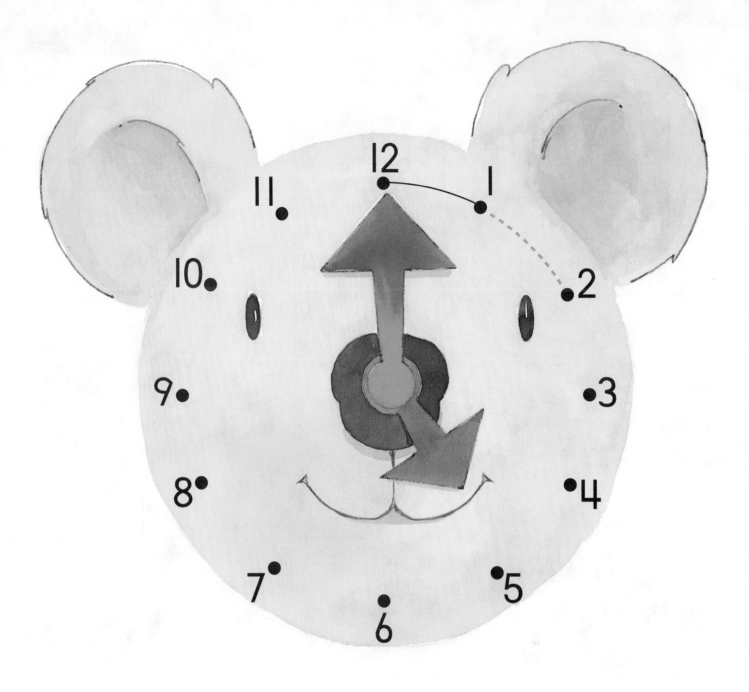

Connect the numbers in order from 1 to 12.

Photography Credits:

Harcourt Brace & Company Photographs

Key: (t) top, (b) bottom, (l) left, (r) right, (c) center.

3 Sheri O'Neal; 5 (t), (c) Rich Franco; (b) Sheri O'Neal; 6 Terry Sinclair; 14, (bl), 15 (tr), 16 (b) Rich Franco.

Illustration Credits:

Elizabeth Allen: cover, 2; **Joe Boddy:** 7, 8; **Jennifer Bolten:** 11, 12; **Teresa Cox:** 3, 4; **Tom Lochray:** 13, 16.

Harcourt Brace School Publishers

Dear Family,
Today we started Chapter 10. We will learn to put things in order as first, next, last and first, second, third, fourth. We will also learn how time is divided into hours, days, weeks, months, and seasons and how a day is divided into morning, afternoon, and night. Here and in the Home Notes are some things you can do to help me learn math.

Love,

_____

Talking Math

Use these math words as you talk with your child about his or her work.

first, second, third, fourth, fifth, sixth, seventh, eighth, ninth, tenth

first, next, last

morning, afternoon, night

clock, hours, minutes, o'clock

seasons: fall, winter, spring, summer

Doing Math

- Use the words of order and time often with your child.

- Have your child mark off each day on a calendar.

- Show your child the hour and minute hands on an analog clock.

- Help your child read the hour on a digital clock.

Playing Math

Find the three pictures that go together. Cut out the numbers. Put them on the three pictures to show the order of what happens. Tell the story.

4

1 2 3

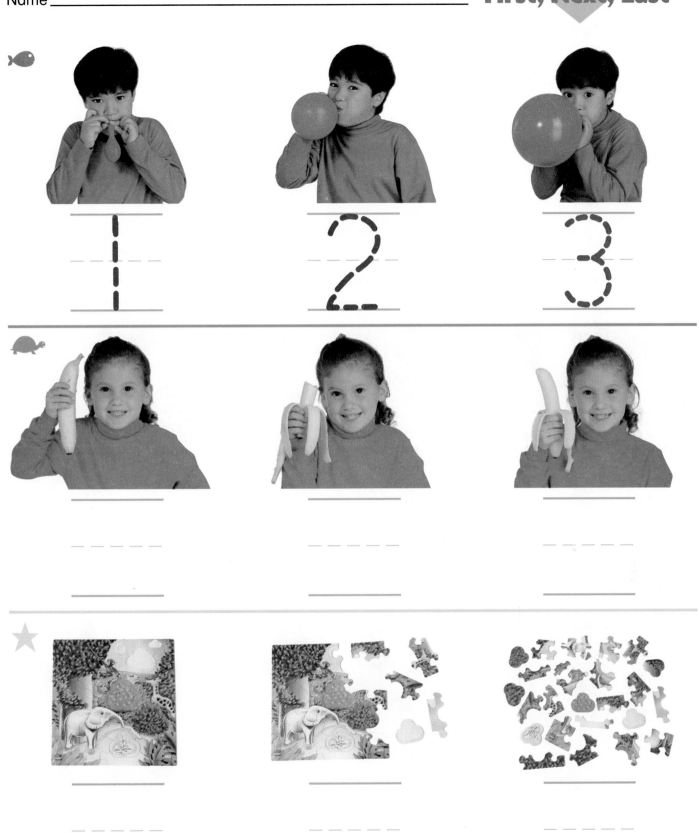

🐟 🐢 ⭐ Write 1, 2, and 3 to show
what happens first, next, and last.

1   3   2

_____   _____   _____

_____   _____   _____

 Write 1, 2, and 3 to show
what happens first, next, and last.

HOME NOTE
Have your child tell a story using the
words *first*, *next*, and *last*.

6

🐟 **Circle the third elephant. Draw a line under the second elephant. Mark an *X* on the fifth elephant.**
🐢 **Circle the eighth bear. Draw a line under the seventh bear. Mark an *X* on the ninth bear.**

Color the first animal red.
Color the fourth animal blue.
Color the tenth animal yellow.

8

HOME NOTE
Have your child use position words such as *first* and *third* to tell the order of the animals in the parade.

_____ o'clock

_____ o'clock

_____ o'clock

_____ o'clock

_____ o'clock

_____ o'clock

Read the time.
Write the number that tells the hour.

Read the time.
Circle the digital clock that shows the
same hour as the clock with hands.

HOME NOTE
Have your child read time on the hour
on real digital and analog clocks.

What Time Is It?

YEAR

MONTH

			1	2	3	4
5	6	7	8	9	10	11
12	13	14	15	16	17	18
19	20	21	22	23	24	25
26	27	28	29	30	31	

By _____

HOME NOTE
Invite your child to share this book with you.
Ask your child the questions on each page.

What do you do at this time?

3

TAKE-HOME BOOK
Help children assemble the book.
Use the questions in the Teacher's Guide to have children
complete the book. Have children take the book home and share
it with family members.

CHAPTER TEN 13

Which happens first?

2

aBbCcDdEeFfGgHhIiJj KkLlM

What time is it?

4

What time is it?

5

What can you do in this season?

7

APRIL

S	M	T	W	T	F	S
		1	2	3	4	
5	6	7		9	10	11
12	13	14	15	16	17	18
19	20	21	22	23	24	25
26	27	28	29	30		

6 **What date is missing?**

What time is it?

Time to go!
Bye!

8

HARCOURT BRACE

MATH ADVANTAGE

CHAPTER
11
Exploring Addition

DIRECTIONS
Write the number that tells how many there are in each group.
Circle the group that has one more.

Photography Credits:
Harcourt Brace & Company Photographs
Key: (t) top, (b) bottom, (l) left, (r) right, (c) center.
Cover, 2 Terry Sinclair; 3 Rich Franco; 9, 10 Victoria Bowmen; 9, 10 Bartlett Digital Photography.
Illustration Credits: Dennis Greenlaw: 3, 4; Bob Holt: 5, 6, 11, 12; Barbara Hranilovich: 7, 8; Holly Cooper: 13, 14, 15, 16.

Harcourt Brace School Publishers

Dear Family,
Today we started Chapter 11. We will learn to add two numbers to find out how many in all. Here and in the Home Notes are some things you can do to help me learn math.

Love,

_____

Talking Math

Use these math words as you talk with your child about his or her work.

join and **in all**
4 dogs joined by 2 dogs is 6 dogs in all.

addition sentence
4 + 2 = 6
4 plus 2 equals 6

Doing Math

- Play a shopping game. Tag some small objects with prices from 1¢ to 5¢. Have your child use pennies to "buy" the objects.

- Let your child help you count out objects such as vegetables, fruits, and crackers. Have your child add two to each group of objects and tell how many objects there are in all.

Playing Math

START → 3¢ 2¢ 1¢ 0¢ 4¢ 3¢ 7¢ 3¢ 5¢ 9¢ 0¢ 4¢ 2¢ 5¢ 8¢ 6¢ 2¢ 7¢ 9¢ 4¢ 5¢ 3¢ 8¢ 6¢ 0¢ 7¢ END

MATERIALS FOR EACH PLAYER: object to use as game marker
Cut out the number cards, and turn them face down.
DIRECTIONS: Players draw a card to tell how many spaces to move. Each player adds one to the number of cents on which he or she lands. If the player adds the numbers correctly, the marker stays on that space. If not, it goes back to where it was.

1   2   3   4

Tell a story about the picture.
Put a connecting cube on each pet. Join the cubes.
Count. Write how many pets there are in all.

_____

- - - - - - - - - -

_____

_____

- - - - - - - - - -

_____

_____

- - - - - - - - - -

_____

_____

- - - - - - - - - -

_____

Tell a story about the picture. Put a connecting cube on each animal. Join the cubes. Count. Write how many animals there are in all.

 HOME NOTE
Have your child draw a picture that shows the joining of two groups.

6

Harcourt Brace School Publishers

Count the bees on the beehive. Then draw one more bee coming. Write how many bees there are in all.

 HOME NOTE

Count the bees on the beehive. Then draw one more bee coming. Write how many bees there are in all.

Tell your child a story that adds one more. Have your child use objects to act out the story.

How Many Pennies?

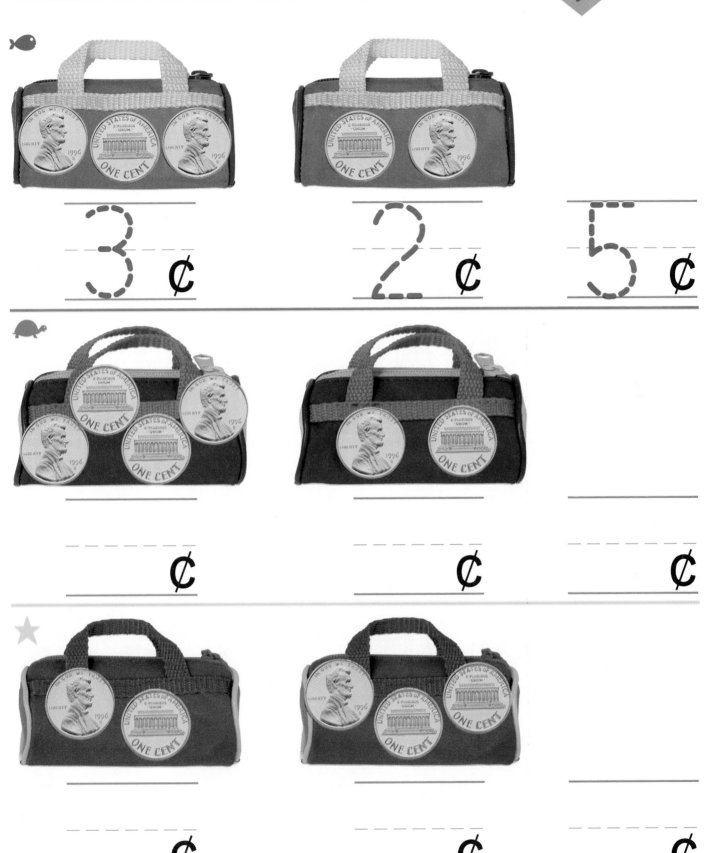

3 ¢   2 ¢   5 ¢

___ ¢   ___ ¢   ___ ¢

___ ¢   ___ ¢   ___ ¢

Write how many pennies are in each purse.
Then write how many there are in all.

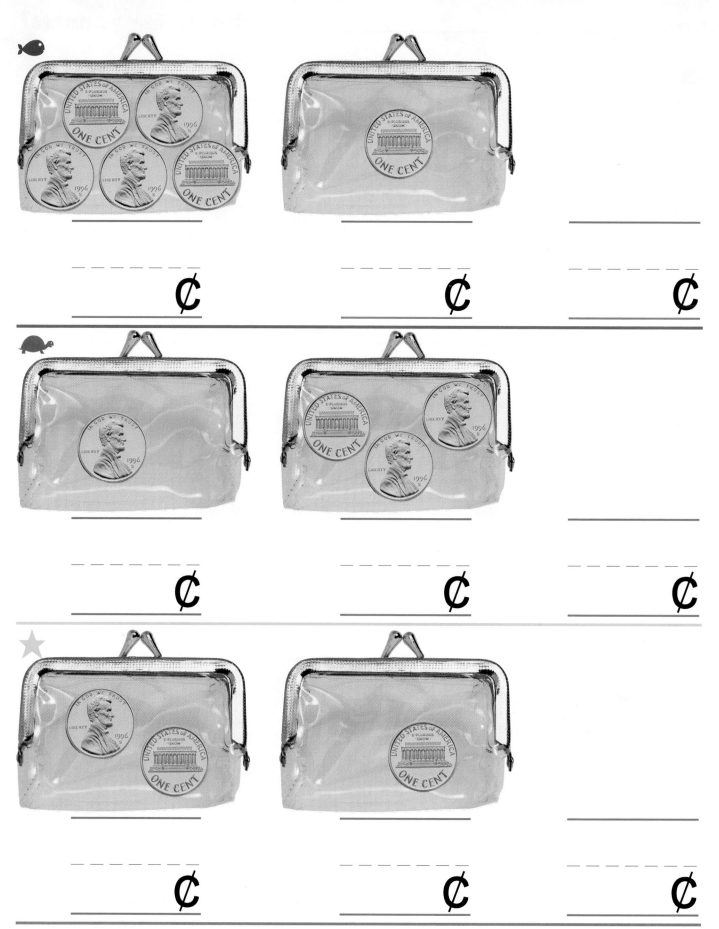

¢   ¢   ¢

¢   ¢   ¢

¢   ¢   ¢

 Write how many pennies are
in each purse. Then write how many there
are in all.

HOME NOTE
Have your child use real coins to
show how many pennies are in
each group.

10

$$3 + 2 = 5$$

$$\_\_\_ + \_\_\_ = \_\_\_$$

$$\_\_\_ + \_\_\_ = \_\_\_$$

Tell a story about the animals in the picture. Write the addition sentence that tells the story.

_____   _____   _____

_____ + _____ = _____

_____   _____   _____

_____   _____   _____

_____ + _____ = _____

_____   _____   _____

_____   _____   _____

_____ + _____ = _____

_____   _____   _____

 Tell a story about the animals in the picture. Write the addition sentence that tells the story.

HOME NOTE
Have your child draw a picture that tells a story about joining groups. Then have him or her write the addition sentence.

12

Going to Bed

By _____

HOME NOTE
Read the story. Then have your child read the addition sentence.

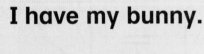

I have my bunny.

3

TAKE-HOME BOOK
Help children assemble the book.
Use the questions in the Teacher's Guide to have children complete the book. Have children take the book home and share it with family members.

CHAPTER ELEVEN   13

I'm going to bed.

2

I have my bunny.
Now I have my teddy.

$$1 + 1 = 2$$

4

14

I have my bunny and my teddy.
Now I have my duck.

2 + 1 = ☐

5

I have my bunny, my teddy,
my duck, and my dog.
Now I have my blanket.

4 + 1 = ☐

7

I have my bunny, my teddy,
and my duck.
Now I have my dog.

$$3 + 1 = \boxed{}$$

6

Now I can go to sleep.

8

16

CHAPTER
12
Exploring
Subtraction

🐟 🐢 ⭐ **Count the cherries. Draw a group that has fewer cherries.**

Photography Credits:

Harcourt Brace & Company Photographs

Key: (t) top, (b) bottom, (l) left, (r) right, (c) center.

Cover Eric Camden; 2 Bartlett Digital Photography; 3 (br) Rich Franco; 9 (tl) Sheri O'Neal; (cl), (bl) Bartlett Digital Photography, 10 Bartlett Digital Photography.

Illustration Credits:

Elizabeth Allen: 7, 8; **Lindy Burnett:** 13, 16; **Patrick Girouard:** 5, 6; **Liisa Chauncy Guida:** 11, 12; **Jean Hirashima:** 3, 4.

Dear Family,
Today we started Chapter 12. We will learn to take away objects from a larger group of objects. We will also learn to write subtraction sentences. Here and in the Home Notes are some things you can do to help me learn math.

Love,

_____

Talking Math

Use these math words when helping your child with subtraction.

take away, are left
There are 3 flowers.
Take away 1 flower.
Now there are 2 flowers left.

subtraction sentence
$3 - 1 = 2$
3 minus 1 equals 2.

Doing Math

- Give your child up to 5 objects. Have him or her take away 1 or 2 objects and tell how many are left.

- Have your child use objects and make up a story in which some of the objects are taken away.

- Have your child draw a picture that shows a subtraction story and write the subtraction sentence that goes with the picture.

Playing Math

MATERIALS: a game marker and 10 pennies for each player
DIRECTIONS: Cut out the number cards and place them face down. A player draws a card, moves that many spaces, reads the number, and puts that many pennies in the wishing well. The player who has more pennies left when he or she gets to END is the winner.

Harcourt Brace School Publishers

2

🐟 🐢 ⭐ ♥ Tell a subtraction story. Put a counter on each animal. Count. Then move counters away to show the animals that are leaving. Write the number that tells how many animals are left.

🐟 🐢 ⭐ ♥ Tell a subtraction story.
Put a counter on each animal. Count.
Then move counters away to show the
animals that are leaving. Write the number
that tells how many animals are left.

HOME NOTE
Have your child tell a subtraction
story about each picture. You may
wish to have your child use objects
to act out the story.

Harcourt Brace School Publishers

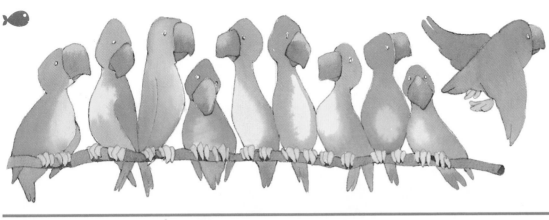

9

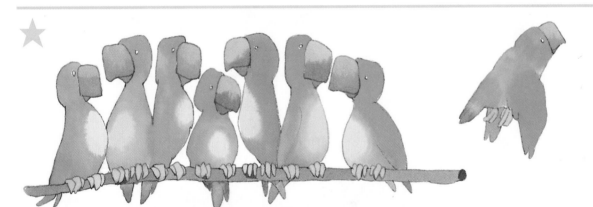

Count the birds. Use your hand to cover the bird that is flying away. Write the number that tells how many birds are left.

 Count the ducks. Use your hand to cover the duck that is turning away. Write the number that tells how many ducks are left.

HOME NOTE
Tell your child a story in which one thing goes away. Have your child use objects to act out the story.

Harcourt Brace School Publishers

$$7¢ - 5¢ = 2¢$$

$$3¢ - 1¢ = \_\_\_ ¢$$

$$6¢ - 3¢ = \_\_\_ ¢$$

Listen to the story. Use pennies to act out the story. Write how many pennies are left.

3¢

$$4¢ - 3¢ = \underline{\quad} ¢$$

4¢

$$6¢ - 4¢ = \underline{\quad} ¢$$

4¢

$$5¢ - 4¢ = \underline{\quad} ¢$$

🐟 🐢 ⭐ **Listen to the story. Use pennies to act out the story. Write how many pennies are left.**

HOME NOTE
Have your child use real pennies to tell each story.

10

Harcourt Brace School Publishers

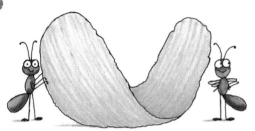  

5 - 3 = 2

___ ___ ___ = ___

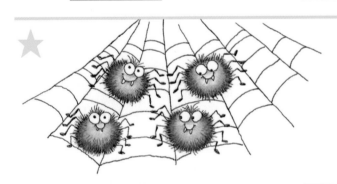 

___ ___ ___ = ___

  ★ **Tell the subtraction story. Then write the subtraction sentence.**

_____ _____ _____

— _____ = _____

_____ _____ _____

_____ _____ _____

— _____ = _____

_____ _____ _____

_____ _____ _____

— _____ = _____

_____ _____ _____

 Tell the subtraction story.
Then write the subtraction sentence.

12

Harcourt Brace School Publishers

HOME NOTE
Have your child tell a subtraction
story about each picture and read the
subtraction sentence that also tells
the story.

By _____

Harcourt Brace School Publishers

HOME NOTE
Read this story to your child.
Have him or her read the subtraction sentences.

I can pretend this pea is an explorer
going into a cave.
How many peas do I have left?

$5 - 1 = \boxed{}$

3

TAKE-HOME BOOK
Help children assemble the book.
Use the questions in the Teacher's Guide to have children
complete the book. Have children take the book home and share
it with family members.

CHAPTER TWELVE **13**

2   **My mom says I have to eat my peas.**

**I can pretend this pea is an astronaut
landing on the moon.**

4   **How many peas do I have left?**

$$4 - 1 = \boxed{}$$

14

I can pretend this pea is a diver
hunting for treasure.
How many peas do I have left?

$$3 - 1 = \boxed{}$$

5

I can pretend this pea is an acrobat
doing a flip.
How many peas do I have left?

$$1 - 1 = \boxed{}$$

7

I can pretend this pea is a mountain climber creeping to the top.
6 How many peas do I have left?

$2 - 1 = \boxed{}$

8 Hey! Look at me! I've eaten all my peas.

16